The R.A.M.S. Library of Alchemy

Volume 9

Summa Perfectionis

by

Geber
(Abu Mūsā Jābir ibn Hayyān)

R.A.M.S. Publishing Company

Summa

Perfectionis

By

Geber

(Abu Mūsā Jābir ibn Hayyān)

Produced by

Restorers of Alchemical Manuscripts Society

R.A.M.S. Publishing Company

R.A.M.S. Publishing Company
117 Rutherford Lane
Stuarts Draft VA 24477

First Edition 2015

ISBN-13 **978-1508670278**

ISBN-10 **1508670277**

Image Processing by Philip N. Wheeler

Printed in the United States of America

GEBER

SUMMA PERFECTIONIS

Collected and Digested by:

WILLIAM SALMON

professor
of
PHYSICK

RAMS
PRODUCTION
1977

Table of Contents

Introduction

Philip N. Wheeler

Summa Perfectionis, "The Height of Perfection," is attributed to Abu Mūsā Jābir ibn Hayyān (c. 721-815), who is most commonly known as Geber. This volume was selected by Hans Nintzel for inclusion in the R.A.M.S. Library due to numerous references to the text in major Alchemical works. Whatever the origin, this text has had a strong influence on the study of Alchemy for many centuries.

Dedicated to Hans W. Nintzel,

American Alchemist

and

Founder of the

Restorers of Alchemical Manuscripts Society

(R.A.M.S.)

Disclaimer

Liability: The publisher does not warrant or assume any legal liability or responsibility for the accuracy, completeness, or usefulness of any information, apparatus, product, or process disclosed. The publisher makes no representation as to the accuracy or completeness of the contents of this book and specifically disclaims any implied warranty of merchantability or fitness for a particular purpose. No warranty may be created or extended by written sales materials or sales representatives. You should obtain professional consultation where appropriate. The publisher shall not be liable for any loss of profit or other commercial or personal damages, including but not limited to special, incidental, consequential, or other damages.

Summa Perfectionis

CHAPTER XXXVII
AN INTRODUCTION INTO THE WHOLE WORK

I. *Perfection* and *Imperfection* of Metalline Bodies, is the Subject of this present discourse; and therefore we treat of things perfecting and corrupting, or destroying, because opposites set near to each other, are the more manifest.

II. That which perfects Imperfect Metals, is a commixion of *Argent Vive* and *Sulphur* in due proportion, by a due and temperate decoction in the bowels of clean, inspissate, and fixed Earth, joyned with an incorruptible radical humidity, whereby it is brough to a solid, fusible substance, with a convenient fire, and made maleable.

III. But Imperfect Minerals are made of a coinmixtion of pure Argent Vive and Sulphur, without due proportion, or a due de coction, in the bowels of the unclean, not fully inspissated, nor fixed Earth, joyned with a corrupting humidity, whereby are brought forth Metals of a porous substance, and though fusible, not sufficiently, or so perfectly maleable as the others.

IV. Under the first definition, are concluded, *Sol* and *Luna,* each according to their perfection: Under the second *Saturn, Jupiter, Mars,* and *Venus,* each according to their imperfection: in which that which is manifest must be hidden, or taken away, and that which is hidden, must be made manifest and brought into operation, which is done by preparing them, by which, their Superfluities will be removed, and their defects, or imperfection supplied, and the true perfection inserted into them.

V. But the perfect Bodies, as Sol and Luna, need none of this preparation they must have, as may subtilize their parts, and reduce them from a Corporality to a fixed Spirituality; that from thence may be made a fixed Spiritual Body, in order to compleat the Great Elixir, whether White or Red.

VI. In both these, viz, the White and Red Elixirs, there is no other thing than Argent Vive and Sulphur, of which one cannot act, not be without the other: It would be a foolish and vain thing to think to make this Great Elixir or Tincture, from anything, in which it is not, this was never the intention of the Philosophers, though they speak many things by similitude.

VII. And because all Metalick Bodies are compounded of Argent Vive and Sulphur, pure, or impure, by accident, and not innate in their first nature, therefore by convenient preparation, 'tis possible to take away their impurity; the end of preparation is to take away Superfluities, and supply the defects.

VIII. For we have considered the substance of Metaline Bodies, perfect and imperfect to be one, viz. Argent Vive and Sulphur, which are pure and clean before their commixtion; and by consideration and experience, we found the Corruption of Imperfect Bodies to be by accident; but that being prepared and cleansed from all their Superfluities, Corruption, and fugitive Uncleanness, we found them of greater brightness, clearness, and purity, than the naturally perfect Metals not prepared, by which consideration we attained to the perfection of this Science.

IX. The Imperfect Bodies have accidentally Superfluous Humidities, and a Combustible Sulphureity, with a Primary Blackness in them and corrupting them; together with an *Unclean, Feculent, Combustible.,* and *very gross Earthiness,* impedeing *Ingress* and *Fusion:* Therefore it behoves us with artificial fire, by the help of purified Salts and Vinegars, to remove superfluous accidents, that the only radical substance of Argent Vive and Sulphur, may remain; which may indeed be done by various ways and methods, according as the Elixir requires.

X. The general way of preparation is this. 1. With fire proportional, the whole superfluous and Corrupt humidity in its essence must be elevated: and the subtil and burning Sulphureity removed; and this by Calcination. 2. The whole Corrupt substance of their superfluous burning humidity and blackness, remaining in their calx, must be corroded with the following cleansed Salts and Vinegars till the Calx be White or Red (according to the nature of the body) and is made clean, and pure from all Superfluity and Corruption: These Calxes are cleansed with the said Salts and Vinegars, by grinding, imbibing and washing. 3. The, whole unclean Earthiness, and Combustible, gross Faeculency, must be taken away with the aforesaid things, not having Metallick Fusion, by commixing and grinding them together with the aforesaid Calx, depurated in the aforesaid manner: For these in the Fusion or Reduction of the Calx, will remain with themselves the said uncleanness and gross Earthiness, the Body remaining pure.

XI. Being thus cleansed, it is Meliorated thus. First, This Purged and Reduced Body is again Calcined by Fire, with the Salts as aforesaid. Secondly, Then with such of these as are Solutive, it must be Dissolved. For this Water is Our Stone, and Argent Vive of Argent Vive, and Sulphur of Sulphur, abstracted from the Spiritual Body, and subtilized or attenuated; which is Meliorated, by confirming the Elemental Virtues in it, with other prepared things of its own kind, which augment the Colour, Fixion Weight, Purity and Fusion, with all other things appertaining to the true Elixir.

XII. The Salts and Vinegars for this work are thus prepared and cleansed. Common Salt, and Salt Gem, as also Sal Alcali, and Sandiver, are cleansed by Calcining them, and then casting them into hot water to be Dissolved, which Solution being Filtred is to be coagulated by a gentle fire, then to be Calcined for a Day and a Night in a moderate fire, and so kept for use.

XIII. Sal Armoniack is cleansed, by Grinding it with a preparation of Common Salt cleansed, and then subliming it in an high Body and Head, till it ascends all pure: then dissolving it in a Porphyrie in the open Air, if you would have it in a water, or otherwise keeping the sublimate in a Glass close stopt for use.

IV. Rock Alums, or Factitious, or other Alums, are

cleansed, by putting them in an Alembick, and extracting their whole Humidity, which is of great use in this Art. The Foeces remaining in the Bottom, Dissolve on a Porphyrie, in a moist place, or in water, and then again extract, and keep it for life.

XV. Vitriol of all kinds is cleansed, by dissolving it in pure Vinegar, then Distilling and Coagulating. Or first abstract its Humidity over a gentle fire: the Foeces Calcine, and Dissolve per deliquium, or in their own water, filtre, and Coagulate (or if you please, the water,) and keep it for use.

XVI. Vinegars of what kind or how acute and sharp soever, are cleansed by subtilization, and their Virtues and Effects are Meliorated by Distillation. With these Salts and Vinegars, the imperfect Bodies may be prepared, purified, meliorated and subtilized, by the help of the Fire. Glass and Borax are pure, and need no preparation.

XVII. Out of the Metalline Bodies we compose the Great Elixir, making One substance of many, yet so permanently fixed, that the strongest or greatest force of Fire cannot hurt it, or make it f lie away, which will mix with Metals in Flux, and flow with them, and enter into them, and be permixed with the fixed substance which is in them, and be fixed with that in them which is incombustible; receiving no hurt by anything which Gold and Silver cannot be hurt by.

XVIII. Hence we define Our Stone, to be agenerating or Fruitful Spirit and Living-water, which we name the Dry-water, by Natural preportion cleansed and United with such Union, that its principles can never be separated one from another; to which two must be added, a third, (for shortning the work) and that is

XIX. The generating or Fruitful Spirit, is White in Occulto, and Red and Black on either side, in the Magistery of this work: but in Manifesto, on both sides tending to Redness. And because the Earthy parts are throughly and in their least particles United with the Airy, Watery, and Fiery, so that in Resolution, no one of them can be separated, but each with all and everyone is dissolved, by reason of the strong Union, which they have with each other in their said least particles, the Compositum is made one solid, uniform substance, the same in Nature, Properties, and all other respects as that of Gold.

Geber's hermetic impress
(from Stolclus's Hermetic Garden).

CHAPTER XXXVIII

OF THE ALCHYMIE OF SULPHUR.

I. SULPHUR is a Fatness of the Earth, thickened by a temperate Decoction in the Mines of the Earth, until it be hardened and made dry, homogeneal, and of an Uniform substance as to its parts. It cannot be Calcined, (without great industry) but with much loss of its substance; nor can it be fixed unless it be first Calcined: but it may be mixed, and its flight in some measure hindred, and its Adustion repressed, and so the more easily Calcined.

II. By Sulphur alone nothing can be done, our work from it alone cannot be perfected, the Magistry would be prolonged even to desperation: but with its Compere (Arsenick for the White, and Antimony for the Red) a Tincture is made, which gives compleat weight to every of the Metals, cleanses and exalts them: and it is perfected without Magistery, without which it performs to us none of these things, but either corrupts or blackens.

III. He who knows how to commix and Unite it amicably with Bodies, knows one of the greatest Secrets of Nature, and one way to perfection: for there are many ways to that Elixir or Tincture. Whatsoever Body is Calcin'd with it receives weight: Copper from it assumes the likeness of Sd. Mercury sublimed with it becomes Cinnabar. All Bodies, except Sol and Jupiter, are easily Calcin'd with it, but Sol most difficultly.

IV. The less Humidity anybody has, the easier it is Calcin'd with Sulphur; it Illuminates everybody, because it is Light, Alum, or Salt, and Tincture. It is difficultly Dissolved, because of its deficiency of Saline parts, but abounding with Oleaginous. It is easily sublimed because of its Spirit; but if it be mixed with Venus, and United to it, it makes a wonderful Violet Colour.

V. That Sulphur is a Fatness of the Earth appears from its easie Liquefaction, and Inflamability, for nothing is inflamed but what is Oleaginous, or melts easily by Heat, but what has such a Nature: yet has it a perfecting middle Nature in it; but this middle substance, is not the cause of the perfection of Bodies, or of Argent Vive, unless it be fixed: 'Tis true, its not easily made to fly; (this he means doubtless of its Spirit or Oyl;) yet it is not perfectly fixed: from whence it is evident, that Sulphur is not the whole perfection of the Magistery, but only a part thereof.

VI. Sulphur commixed with Bodies, burns, some more, others less; and some resist its combustion, and some not; by which may be known the difference between those Bodies which are wanting in perfection, tho' prepared for the great work. Sol is not easily to be burned by Sulphur: The next to this is Jupiter, then Luna, after that Saturn, then Venus, that is more easily burnt, which is farther distant from the Nature of the Perfect.

VII. Also from what Radix the imperfect Body proceeded or was generated, it appears from the diversity of Colours after Combustion: Thus Luna obtains a black mixt black mixt with Azure: Jupiter, a black mixt with a little Redness: Saturn a dull black, with much Redness and a Livid Colour: Venus, a black with a Livid; if it be much burnt, if but a little, a pleasant Violet: Mars, a black dull Colour. But if Sulphur be commixt with Sol, he obtains an Intense Citrine Colour.

VIII. Sol and Luna Calcin'd with Sulphur, being reduced, return into the Nature of their own proper Bodies. Jupiter, Calcin'd and reduced, recedes its greater part: Saturn has sometimes a greater, sometimes a lesser part destroyed. But Saturn and Jupiter are both preserved, by a right and gentle Reduction, yet they rather tend to another Body than their own, as Saturn into a dull Coloured (Regulus of) Antimony, Jupiter into a bright Coloured (Regulus of) Antimony. Venus is diminished in the Impressions of Fire in her reduction, but withal ponderous, augmented in weight, soft, of a dull Citrine Colour, partaking of blackness: And Mars is more diminished in the Impression of the Fire than Venus; by which things are found out, the Nature of all Bodies that are altered.

IX. *The Preparation of Sulphur. 1. Take the best Green Sulphur Vive, Grind it to a subtil Pouder, Boyl it in a Lixivium of Pot-Ashes and Quicklime, gathering from the Supersities its Oyliness, till it, appears to be clear. Stir the whole with a Stick, and immediately decant the Lixivium with the pure parts of the Sulphur, leaving the more gross parts behind: let the Liquor cool, and pour upon it a fourth part of the quantity, of Spirit of Vinegar; so will a white Pouder precipitate, white as Milk, which dry with a gentle Heat, and keep for use.*

X. 2. *Take of this prepared white Sulphur; Scales of Iron Calcin'd to Redness, Roch—Alum well Calcin'd ana One Pound, Common Salt prepared, Half a Pound: Incorporate all these well by Grinding them together with Vinegar, that the whole may be Liquid, which then boil, stirring it till it be all very black: then dry and grind to a fine Pouder, which put into an Aludel of a Foot and half high, with a large Cover; and let the Cover of the Alembick have a broad Zone or Girdle, for Conservation of the Spirits elevated, then sublime according to Art; the light Flos which adheres to the sides of the Alembick, cast away, for it is combustible, defiled, and defiling. But the close, compact, or dense Matter sublimed in the Zone, put by itself into a Phial, and Decoct it upon an Ash Heat, so long till its Combustible Humidity be exterminated, then keep it in a clean Vessel for use:* Note, that Sulphur and Arsenick sublimed from the Calx of Copper, are more whitened, than when sublimed from the Calx of Iron.

Metamorphoses of Prime Matter in the alchemical "Vessel."
The salamander symbolizes the igneous principle, "Philosophic Sulphur."

CHAPTER XXXIX
OF THE ALCHYMIE OF ARSENICK.

I. ARSENICK, is also a fatness of the Earth, as is afore declared of Sulphur, having an inflamable substance, and a subtil matter like to Sulphur; but it is diversified from Sulphur in this, viz. That it is easily made a Tincture of Whiteness, but of Redness with great difficulty; whereas Sulphur is easymade a Tincture of Redness, but of Whiteness, most difficultly.

II. Of Arsenick, there is a Citrine and a Red, which are profitable in this art, but the other kinds not so: Arsenick is fixed as Sulphur, but the sublimation of either is best from the Calx of Metals: But neither Sulphur nor Arsenick, are the perfective matter of this Work, they not being compleat to perfection, though they may be a help to perfection, as they may be used. The best kinds of Arsenick, are the Sciffile, the Lucid, and Scaly.

III. This Mineral also (like as Sulphur) has a perfecting middle Nature in it, which yet is not the cause of the perfection of Bodies, or of Argent Vive; unless it be fixed; but being fixed, this Spirit is an agent of the White Tincture: What we have said of

Sulphur in the former Chapter, at Sect. 5 may be understood here.

IV. Because in Arsenick the Radix of its Minera, in the action of Nature has many inflamable parts of it resolved, therefore the work of its separation is easie, this being the Tincture of Whiteness, as Sulphur is of Redness.

V. To prepare Arsenick. Being beaten into fine pouder, it must be boiled in Vinegar, and all its combustible fatness extracted as in Sulphur, Chap. 38. Section 9. Then take *of the prepared Arsenick Copper calcin'd ana one Pound: Alum calcin'd, common Salt prepared, ana half a Pound: Having ground them well together, moisten the mixture with Spirit of Vinegar, that it may be liquid, and boil the same, as you did in the Sulphur: Then sublime it in an Aludel, with an Alembick, of the heighth of one foot: what ascends white, dense, clear, and lucid, gather and keep it, (as sufficiently prepared) for the use of the Work.*

VI. Or thus: Take of Arsenick prepared by boiling, filings of Copper, ana one Pound: Common Salt, half a Pound: Alum calcined four Ounces; grind them exactly with Spirit of Vinegar, then moisten till they be liquid, and stir them over a fire till the whole be blackened. Again, Imbibe and dry, stirring as before, do this a third time, then sublime as above directed.

VII. To fix Arsenick and Sulphur. They are fixed two ways, viz. 1. By manifold Sublimations. 2. By precipitation of them sublimed into heat. The first way, Reiterate their Sublimations in the Vessel Aludel, till they remain fixed. This Reiteration is made by two Aludels, with their two Heads, or Covers in the following order, that you may never cease from the Work of Sublimation, until you have fixed them. Therefore to soon as they have ascended into one Vessel, put them into the other, and so do continually, never suffering them long to abide, adhering to the sides of either Vessel, but constantly keep them in the elevation of fire, till they cease to sublime.

VIII. The second way. This is by praecipitating it sublimed into heat, that it may constantly abide therein, until it be fixed: and this is done by a long glass Vessel, the bottom of it (made of Earth not of Glass, because that would crack) must be artificially joyned with good luting; and the ascending matter, when it adheres to the sides of the Vessel, must with a Spatula of Iron, or Stone, be put down to the heat of the bottom, and this precipitation repeated, till the whole be fixed.

IX. To sublime Arsenick. Take Arsenick, filings of Venus ana one Pound, Common Salt half a Pound: Alum Calcin'd four Ounces, mortifie with Vinegar, stirring over a fire till all be black:

Again, Imbibe and dry, stirring as before, which repeat again; *the sublime, and it will be profitable.*

CHAPTER XL

OF THE ALCHYMIE OF THE MARCHASITE.

I. The MARCHASITE is sublimed two ways, 1. Without Ignition. 2. With Ignition, because it has a two-fold substance, viz. One pure Sulphur, and Argent Vive mortified. The first is profitable as Sulphur; the second as Argent Vive mortified, and moderately prepared. Therefore we take in this last, because by it we are excused from the former Argent Vive, and the labour of mortifying it.

II. The intire way of the sublimation of this Mineral is, by grinding it to pouder, and putting it into an Aludel, subliming its Sulphur without Ignition; always and very often removing what is sublimed. Then augment the force of the fire into Ignition of the Aludel. The first sublimation must be made in a Vessel of Sublimation, and so long continued, till the Sulphur is separated; the process being successively and orderly continued, until it is manifest that it has lost all its Sulphur.

III. Which may be known thus: When its whole Sulphur shall be sublimed, you will see the colour thereof changed into a most pure White, mixt with a very clear, pleasant, and coelestine colour: Also you may know it thus: Because if it has any Sulphur in it, it will burn and flame like Sulphur; but what shall be secondly sublimed after that sublimate, will neither be inflamed, nor shew any properties of Sulphur, but of Argent Vive mortified, in the reiteration of sublimation.

IV. You must get a solid, strong, well baked Earthen Vessel, about three foot high, but in breadth Diametrically no more than that a hand may commodiously enter: The bottom of this Vessel, (which must be made so that it may be separated and conjoyned, must be made after the form of a plain wooden Dish, but very deep, viz, from its brim to the bottom about seven or eight Inches; from that place, or moveable bottom to the head, the Vessel must be very thickly and accurately glazed within: Upon the head of the Vessel must be fitted an Alembick, with a wide Beak or Nose: Joyn the bottom to the middle, with good tenacious lute (the Marchasite being within that bottom) then set on the Alembick, and place it in a Fornace, where you may give as strong fire, as for the fusion of Silver or Copper.

V. The top of the Fornace must be fixed with a flat Hoop, or Ring of Iron, having a hole in its middle, fitted to the greatness of the Vessel, that the Vessel may stand fast within it: Then lute the junctures in the circuit of the Vessel and the Fornace, lest the fire passing out there, should hinder the adherency of the subliming flowers, leaving only four small holes, which may be opened or shut in the flat Ring or Hoop aforesaid, through which Coals may be put in round about the sides of the Fornace: Likewise four other holes must be left under them, and between their spaces for the putting in of Coals, and six or eight lesser holes, proportionate to the magnitude of ones little finger, which must never be shut, that thereby the fire may burn clear: Let these holes be just below the juncture of the Fornace, with the said Iron Hoop.

VI.   That Fornace is of great heat, the sides of which are to the height of two Cubits, and in the midst whereof is a Round, Grate, or Wheel filled full of very many small holes close together, (wide below or underneath, but small above, or in the superior part,) and strongly annexed to the Fornace by luting, that the Ashes or Coals may the more freely fall away from them, and the said Grate be continually open for the more free reception of the air, which mightily augments the heat of the fire.

VII. The Vessel is of the aforesaid length, that the Fumes ascending may find a cool place and adhere to the sides, otherwise was it short, the whole Vessels would be almost of an equal heat, whereby the sublimate would fly away, and be lost. It is also Glased well within, that the Fumes may not peirce its Pores and so be lost; but the Bottom which stands in the fire is not to be Glazed, for that the Fire would melt it; nor unglazed would the matter go through it, for that the Fire makes it rather to ascend.

VIII. Not let your Fire be continued under your Vessel, till you know that the whole matter is ascended into flowers, which you may prove by putting in a Rod of Earth well burned with a Hole in its end, through a Hole in the Head, about the bigness of ones little Finger, putting it down almost to the middle there, or nigh the matter from whence the sublimate is raised; and if anything ascends and adheres to the Hole in the Rod, the whole matter is not sublimed, but if not, the sublimation is ended.

IX. That the Marchasite consists of Sulphur and Argent Vive, it is sufficiently evident; for if it be put into the fire, it is no sooner Red—Hot, but it is Inflamed and burns: also if mixed with Venus, it gives it the Whiteness of pure Silver; so also if mixed with Argent Vive, and in its sublimation it yields a Coelestial Color, with a Metalick Lucidity.

X. *To prepare the Marchasite. Take the fine Pouder of the Mineral, spread it an Inch thick over the*

32

Bottom of a large Aludel, and gather the Sulphur with a gentle fire. When that is ascended; take off the Head or Alembick, and having applied another, augment the Fire, then that which has the place of Argent Vive Ascends, as we have before declared.

CHAPTER XLI

OF THE ALCHYMIE OF MAGNESIA, TUTIA AND OTHER MINERALS.

I. The Sublimation of MAGNESIA and TUTIA is the same with that of the Marchasite, for that they cannot be sublimed without Ignition, having the same cause, the same Operation, and the same General method: Likewise all imperfect Bodies, are sublimed in the same order, without any difference, except that the Bodies of the Metals must have a more vehement fire than the Marchasite, Magnesia and Tutia: nor is there any diversity in Metaline sublimation save, that some need the addition of some other substance to make them sublime or rise.

II. But in the sublimation of Imperfect Metaline Bodies, no great quantity of the Body to be sublimed, must be at once put into the bottom of the Vessel, because much Metaline substance, holds the parts faster, and hinders the subliming: also the bottom of the sublimatory should be flat, not Concave, that the Body equally and thinly spread upon the bottom, may the more easily sublime in all its parts.

III. Such Bodies as need the admixtion of other

substances, are Venus and Mars, by reason of the slowness of their fusion: Venus needs Tutia; and Mars Arsenick, and with these they are easily sublimed, for that they well agree with them. Therefore their sublimation is to be made as in Tutia, and other like things, and to be performed in the same method and order, as in the former Chapter.

IV. Now Magnesia has a more Turbid and Fixed, and less inflamable Sulphur, and a more Earthy and faeculent Argent Vive, than the Marchasite, and therefore the more approximate to the Nature of Mars.

V. But Tutia is the fume of White Bodies; for the Fume of Jupiter and Venus adhering to the sides of the Fornaces where these Metals are wrought, does the same thing that Tutia does: and what a metalick Fume does not, without the admixtion of some other Body, neither will this likewise do.

VI. And by reason of its subtilty, it more penetrates the profundity of a Metaline Body, and alters it more than it does its own Body, and adhears more in the Examen, as by experience you may find: and whatever Bodies are altered by Sulphur of Argent Vive, will also necessarily be altered by this, because of their Unity in Nature.

VII. To prepare Tutia, Pouder it very fine, and put it into and Aludel and by strong Ignition, or help of vehement fire, cause the Flowers to ascend or sublime, so is it prepared for use. It is also dissolved in Spirit of Vinegar, having been first Calcin'd and so it is also well prepared.

VIII. Also it is certain, that many necessary things for our purpose, are extracted from Imperfect Bodies, which need yet a farther preparation, as first Ceruse; which is thus prepared: Wash it in Spirit of Vinegar, and separate it from its more gross parts; and the Milk coagulate in the Sun, and it is prepared.

IX. Spanish White, Tin, Putty and Minium, are prepared after the same manner, by dissolving them in Spirit of Urine, and then filterating and coagulating in the Sun as before.

X. Verdegrise is dissolved in Spirit of Vinegar, and rubified, being gently congealed, with the soft heat of a gentle fire; and then it is prepared, and made fit for the Work.

XI. Crocus Martis is dissolved in Spirit of Vinegar, and filtred: This Red Water being congealed, yields an excellent Crocus fit for use.

XII. Aes Ustum, or Copper calcin'd, is to be ground to pouder, and washed with Spirit of Vinegar, after the same manner as we taught in the preparation of Ceruse: So in like manner Litharge of Gold and Silver: You may also dissolve these things again, and they will be purer: You may also use them either dissolved or congealed; this is a profound Investigation.

XIII. Antimony is Calcined, Dissolved, Filtred, Congealed, and ground to pouder, and so it is prepared.

XIV. Cinnabar must be sublimed from Common Salt once, and so it is well prepared for use.

XV. The fixation of Marchasite, Magnesia and Tutia. You must after the first sublimation of them is finished, cast away their foeces; and then reiterate their sublimation, so often returning what sublimes to that which remains below -f[1] either of them, till they be fixed, which must be done in proper subliming Vessels.

---

[1] This is exactly as Hans typed. -pnw

GEBER'S FORNACE

Geber. lib. 2. cap. 40.
44

CHAPTER XLII

OF THE ALCHYMIE OF SATURN.

I. To *prepare* LEAD. Set it in a Fornace of Calcination, stirring it while it is in Flux, with an Iron Spatula full of Holes, and drawing off the scum, till it be converted into a most fine pouder: Sift it, and set it in the Fire of Calcination, till its fugitive and inflamable substance be abolished: Then take out this Red Calx, imbibe, and grind it often with Common Salt cleansed, Vitriol purified, and most sharp Vinegar, which are the things to be used for the Red; but for the White, Common Salt, Common Alum, and Vinegar.

II.   Your matter must be often imbibed, dryed, and ground, till by the benefit of the aforesaid things, the uncleanness be totally removed: Then mix Glass therewith, and cause the pure body to descend, that descending (by means of a vehement heat) the pure body may be reduced.

III. Calcine it again with pure Sal Armoniack (as you do Jupiter) and most subtily grind and dissolve it by the way aforesaid, for this is the water of Argent Vive and Sulphur proportionally made, which we use in the Composition of the Red Elixir.

IV. Lead is a Metalick Body, livid, earthy, ponderous, mute, partaking of a little Whiteness, with much paleness, refusing the Cineritium, and Cement, easily extensible in all its dimensions, with small Compression, and very fusible without Ignition. Yet some Men say, that Lead in its own Nature, is much approximated to Gold; these judge of things, not as they are in themselves, but according to sense, being void of Reason, and not conceiving the Truth.

V. It has much of an Earthy substance, and therefore is washed, and by a Lavament converted into Tin, by which it appears, that Tin is more assimilated to the perfect. It is also by Calcination made Minimum; and by hanging over the Vapour of Vinegar, it is made Ceruse. And tho it is not near to perfection, yet by our Art, we easily convert it into Silver, not keeping its Weight in transmutation, but acquiring a new Weight, which it obtains by our Magistry. It is also the Tryal of Silver in the Cupel, as we shall hereafter shew.

VI. It differs not from Tin, after repeating its Calcination to the reduction thereof, save, that it has a more unclean substance commixed of a more grose Sulphur, and Argent Vive, the Sulphur being more burning and adhaesive to the Argent Vive. It has a

40

greater Earthy Faeculency than Jupiter, which appears by washing of it with Argent Vive; and more Faeculency comes from it by washing than from Jupiter, and its first Calcination is easier performed than in Tin, because of its Earthiness: and because its foulness is not rectified as in Jupiter, by repeated Calcinations, it is a sign of greater impurity in its principles, and in its own Nature.

VII. Its Sulphur is not separated from it in fume, but is of a Citrine Colour, of much Yellowness, the like of which is remaining below at the bottom, which shews that it has much of a Combustible Sulphur in it, and because the Odour of Sulphureity is not removed from it in a short time, it shews that it approaches to the Nature of fixed Sulphur, and is Uniformly commixed with the substance of Argent Vive. Therefore when the fume ascends, it ascends with the Sulphur not burning, whose property is to create Citrinity.

VIII. And that the quantity of its not burning Sulphur is more than in Tin, appears for that its whole Colour is changed into Citrinity, in Calcination, but of Tin into White: Whence the cause appear why Jupiter in Calcination is more easily changed into a hard Body than Saturn: the burning Sulphureity being more easily removed from Jupiter than Saturn, one of the causes of its softness is removed; whence (being Calcined) it necessarily follows it must be hardened: but Saturn, because it has both the causes of softness strongly conjoyned, viz, much burning Sulphur and much Argent Vive, it is not easily hardened.

IX. Bodies having much Argent Vive, have much of Extension, but such as have little Argent Vive, have little Extension. Thus Jupiter is more easily and subtily extended than Saturn. Saturn more easily than Venus. Venus more easily than Mars. Luna more subtily than Jupiter. And Sol more subtily than Luna.

X. The Cause of Induration or hardening is fixed Argent Vive, or fixed Sulphur; but the cause of softness is Opposite. The cause of Fusion is also twofold to wit, of Sulphur not fixed, and Argent Vive of what kind soever; Sulphur not fixed is necessarily a cause of Fusion without Ignition. This is evident in Arsenick, for projected on Bodies difficult to be Fused, it makes them of easie Fusion, without Ignition: and the cause of Fusion with Ignition is fixed Argent Vive. But the Impediment of Fusion is fixed Sulphur.

XI. From hence it appears, That seeing Bodies of greatest perfection, contain the greatest quantity of Argent Vive: Those Imperfect Bodies holding more of Argent Vive, must needs be more approximate to the

perfect whence it follows that Bodies of much
Sulphureity, are Bodies of much Corruption.

XII. From hence it is evident, that Jupiter is near
to the perfect, seeing it participates more of
Perfection, but Saturn less; Venus yet less, and Mars
least of all. And as to the Medicines, compleating
them, it is clear, that Venus is the most perfective
of Medicine; Mars less, Jupiter yet less; and Saturn
least of all.

XIII. Thus according to the diversity of Bodies,
diversity of Medicines are found out: A hard Body,
that can endure Ignition requires one Medicine; but
the soft, that abides not Ignition another; that one
may be softned and attenuated in its profundity, and
equalized in its substance; but the other hardned, and
its occult parts inspissated.

XIV. There are three degrees which the Imperfect
Bodies, chiefly Saturn and Jupiter must obtain, in
order to perfection: First, Cleanness, or Brightness:
Secondly, Hardness, or Dens-ness, with Ignition in
fusion. Thirdly, Fixation, by taking away their
fugitive substance.

COITUS.

Pater eius Sol.

Mater eius Luna.

The union of the two principles, Male and Female, Sun and Moon, Sulphur and Mercury, Fire and Water, etc.

44

XV.   They are cleansed (viz. Saturn and Jupiter) in a threefold manner: 1. By Mundifying. 2. By Calcination and Reduction: 3. By Solution. First, By things purifying they are cleansed two ways, either by reducing them into a Calx, or into the Nature of Bodies: reducing into a Calx, they are purified either by Salts, or Alum, or Glass: Thus, when the Body is Calcin'd put upon its Calx, water of Alums, or Salts, or Glass mixed with it, and reduce it to a Body, which so often reiterate till they look purely clean: For seeing Aluins, Salts, and Glass, are fused with another kind of fusion than Bodies, therefore they are separated from them, retaining with themselves the earthy substance, the purity of the Bodies being only left.

XVI. Or thus. Let Saturn or Jupiter be filed, and mix therewith Alums, Salts, and Glass, and then reduced into a body, and this so often to be repeated till they be well cleansed: They are also cleansed by way of Lavement with Argent Vive, of which we have spoken before.

XVII. The second way of cleansing Saturn and Jupiter, by Calcination and Reduction with sufficient fire, whereby they are freed from a twofold corrupting, substance. 1. One inflamable and fugitive. 2. Another earthy and faeculent; because the Fire elevates and consumes every fugitive substance. And by reduction the same fire, divides every substance of earth, with its proportion: See Sect. 1, 2, 3. above.

XVIII. The third way of cleansing Saturn and Jupiter by Solution of their substance, and by reduction of that likewise, which is dissolved from them; for that solution reduced makes them more clean, than any other way or kind of preparation whatsoever, except that by Sublimation, to which this is equivalent.

XIX. Induration, or hardning of their soft substance. This is done with Ignition in their Fusion, thus. With Saturn or Jupiter the substance of Argent Vive, or Sulphur fixed, or of Arsenick, must be mixed in their profundity. Or, they must be mixed with hard, and not fusible things, as the Calx of Marchasite, and Tutia, for these are united with, and embraced by them, and harden them so, that they flow not, till they are red hot. The same thing is also compleated by our Medicine perfecting them, of which hereafter.

XX. Fixation, by removal of their fugitive substance. This is done by calcination in a fire proportional to their substance: In order to which, 1. All their corrupting adustive substance must be cleansed from them as aforesaid. 2. Then their earthy superfluity must be taken away. 3. They must be

dissolved and Reduced, or compleatly washed in a
Lavement of Argent Vive. This is necessary and
profitable.

XXI. Saturn is specially hardened by a Calcination
with the Acuity of Salt, and by Talk it is especially
dealbated, as also by Marchasite and Tutia. Calcine
Saturn fluxed with common Salt putrefied, stirring it
continually with an Iron Spatula, till it comes to
Ashes. Decoct it for one Natural day, and let it be a
little Fiery hot, but not much; then wash it with pure
clean water, and Calcine it for 3 daies till it be Red
both within, and without. If you would have it to be
prepared for the White, Imbibe it with water of White
Alum, and reduce it with Oyl of Tartar, or its Salt.
But if you would have it for the Red, Imbibe it with
the water of Crocus Martis, and of Verdigrise, and
reduce it with Salt of Tartar as before: This work
Reiterate as often as need requires.

XXII. The Calcination of Saturn and Jupiter. Let a
great Test (or Calcining Pan) be placed in a Fornace,
and put Saturn and Jupiter into it, with as much
common Salt prepared, and Roch Alum Calcined: being in
Flux, let the Metal be continually stirred with an
Iron Spatula full of holes, till the whole be reduced
to Ashes, which sift, and set them in the Fire again,
keeping them continually Red Fire Hot till the Calx of

Jupiter is whitened or that of Jupiter is rubified as Minimum.

XXIII. *The Regimen of Saturn and Jupiter for the White. Take Saturn purified three Pound, melt or add to it clean or purified Mercury twelve Pound, stirring the whole that they may be mixed: This mixture put into a Bolt-Head of a Foot in Length, which place in the Athanor with a gentle Fire for a week. Take purified Jupiter one Pound, melt and add purified Mercury 12 Pound, doing in all respects as before with Saturn.* In this weeks time you will have a Paste dissolved, fit to be Fermented with the White Ferment, Thus.

XXIV. Take *of the White Ferment one Pound, of the Paste of Saturn two Pounds, of the Paste of Jupiter three Pounds: These being dissolved, mix through their least pasts, and set in putrefaction, (in a moderate Fire, like as in dissolution) for seven daies: Then take them out well mixed and Strain or Squeeze their more Liquid parts through a Cloth: The thick Matter remaining, put into a Glass, Seal it well up, and place it in an Athanor for the time aforesaid, which do thrice, till it has Imbibed all the humidity. Then put the Vessel with its Matter into a Fornace of Fixation for twelve daies, which done, take it forth, and reduce it with things reducing;* so will you find

that which our Ancestors found not without great
Study, viz. The Generated, generating. Prove this upon
the Cineritium or Cupel with Lead, and you will find
the Body perfect in Whiteness, perpetually generating
its like.

XXV. The Regimen of Saturn is also compleated, if
being prepared and dissolved, (I suppose he means in
his dissolutive Water, made of Nitre and Vitriol) it
be mixed with a third part of its Red Ferment
dissolved also; and then Distilling off the Water, and
Cohobating seven times. Reduce it to a Body, and prove
it by its Examen, and you will rejoyce in the
bountiful Body which is generated.

XXVI. White Medicines for Saturn: also solar
Medicines for Saturn. Because the Medicines, and the
work are wholly or altogether the same, as for
Jupiter; and that in the Chapter of Jupiter we have
explicitely and largely declared the matter, we shall
refer you thither, saying no more thereof in this
place. See Chapter 43 Sect. 16, 17, 18, 19, 20, 21.
following.

CHAPTER XLIII
THE ALCHYMIE OF JUPITER

I. To prepare JUPITER. Put it into a fit Vessel, in a Fornace of Calcinatiori, and make a good Fusion, stirring the melted Metal with an Iron Spatula full of holes, drawing off the Scum as it arises, and again stirring the Body, and thus continuing till the whole quantity is reduced to Pouder or Ashes. This Pouder sift, and replace it in the Fornace again in the same heat of Fire, stirring it often, for 24 hours, till its whole accidental and superfluous humidity is abolished with all its combustible and corrupting Sulphur. Then often well wash it with common Salt cleansed, and Alum purified, and sharp Vinegar, and dry it in the Sun or Air, Grind it again, washing and drying it; doing this so long till by the acuity of the Salts, Alums, Vinegar, its whole humidity, blackness and uncleanness is taken away. This done, add Glass in fine Pouder to it, impaste the whole together, and with a sufficient Fire make it flow in a Crucible with a hole in its bottom, set within another, so will the pure and clean Body descend, the whole Earthy and Faeculent substance remaining above with the Glass, Salts, and Alums; in which pure Body is an equal and perfect proportion Argent Vive, and White Sulphur not burning. Then Calcine this pure Body with pure and clean Sal Armoniack; till it be in weight, equal or thereabout; being well and perfectly Calcined, Grind the whole well and long upon a Porphyrie, and set it in the open Air in a Cold moist place; or in a Glass Vessel in a Fornace of Solution, or in Horse-Dung, till the whole be dissolved, augmenting the Salt if need be. This Water ought to be esteemed, for it is what we seek for in the whole.

II. Tin, is a Metallick Body, White, Livid, not pure, and a little Earthiness, possessing in its Root harshness, softness, easiness of Liquefaction without Ignition not abiding the Cupel or Cement, but extensible under the Hammer. Therefore Jupiter among Bodies diminished from perfection, is in the Radix of its Nature of affinity to Sol and Luna, but more to Luna, and less to Sol.

III. Jupiter, because it receives much whiteness from the Radix of its generation, it whitens all other Bodies which are not White, but it has a fault, that it breaks or makes brittle all other Bodies, except Saturn and most pure Sol: Jupiter adheres much to Sol and Luna, and therefore does not easily receede from them in the examen or Tryal by the Cupel. The Magistery of this Art, gives it a Tincture of Redness, that shines in it with inestimable brightness: It is hardened and cleansed more easily than Saturn. He who knows how to take away its Vice of breaking, will suddenly reap the Fruit of his Labour with joy, because it agrees so well with Sol and Luna, and will never be separated from them.

IV.   In Calcining Tin, a Sulphureous stink arises, from its Sulphur not fixed; and tho it gives no flame, yet it is not fixed, for its not flaming is by reason of the great abundancy of its Argent Vive, preserving from Combustion: So that in Tin is a two-fold Sulphur, and a two—fold Argent Vive: one Sulphur less fixed, sending forth a stink; the other more fixed, because it abides with the Calx in the Fire and stinks not.

V. There is also a two-fold substance of Argent Vive in it, one not fixed, and the other fixed: because it makes a Crashing noise before its Calcination, but after it has been thrice Calcined, that Crashing ceases, which is caused by its fugitive Argent Vive being flown away. This is evident in Lead being wash't with Argent Vive, and then melted in a very gentle fire, some part of the Mercury will remain with the Lead, you will give to it this stridor, converting the Lead into Tin.

VI.   On the contrary also, Tin may be converted into Lead: For by a manifold repitition of its Calcination, and a fire fit for its reduction, it is turned into Lead; but especially when by subtraction of its Scoria, it is calcin'd with a great fire.

VII. Now after the removal of these two Substances, viz. Sulphur, and Argent Vive from Jupiter, you will find that it is livid, and weighty as Lead, yet partaking of greater whiteness than Lead, and therefore more pure than Lead: In which is the equality of fixation, of the two compounding things, viz. Sulphur and Argent Vive, but not the equality of quantity, because in the Commixtion, the Argent Vive, is super—eminent.

VIII. Now if there were not in its proper nature a greater quantity of Argent Vive than of Sulphur, Argent Vive would not easily adhere to it: For which reason it adheres with difficulty to Venus; but with much greater difficulty to Mars, by reason of the small quantity of Argent Vive contained therein; the sign of which, is the easie fusion of the one, and the difficult fusion of the other.

IX. But the fixation of these two substances remaining, approaches nigh to firm fixation, yet it is not absolutely fixed, which is evident from the calcination of its body, and after calcination, the exposing the same to the most strong fire; for by that, division is not made, but the whole purified, from whence it appears, that the burning Sulphur in

Tin, is more easily separated than that in Lead: And that, because its corrupting Properties are not radical, but accidental, therefore they are the more easily separated, and its mundification, Induration, and fixation, the more speedy.

X. And because, that after Calcination and Reduction, we found in its fume a Citrinity, through the great force of fire; we judged, that it contained in its body much fixed Sulphur: By these Operations you may find out the Principles of Bodies, and the Properties of Spirits.

XI. At Sect. 14, 15, 16, 17, 18, 19, 20, 21, 22. of the former Chapter, we have shewn the farther preparations of Tin, which because they are so plainly expressed there, are needless to be repeated here again. Yet there are other special Preparations which are the following, to wit, by Calcination, by which its substance is more hardened, which happens not to Saturn. Also, by Alums, for these properly harden Jupiter. Also, by Conservation of it in the fire of its Calcination, for by this it loose its stridor or Grashing, and fraction of bodies likewise, the which in like manner happens not to Saturn.

XII. Calcine Jupiter (as Saturn at Sect. 21 of the former Chapter with Common Salt purified) and whiten its Calx for three days as in Saturn: But see you err not in its Reduction, for that is difficult unless it be made in the Fornace, by Citteritium or Cement; then it is done with ease. But that you may not err, joyn that Body which you would reduce, in equal parts with that by which you make the reduction, and co-unite the divided Calx: But in Tinctures there is another consideration, for the matter tinging must be multiplied upon the matter to be tinged, till the Tincture appear in the Body or Medicine.

XIII. After you have found these two Leads, and found their color and brightness, with other things according to your desire; possibly they may yet want Ignition; then you must thus proceed. Dissolve Tutia calcined, and Tin calcined, mix both Solutions, and with that water imbibe the Calx of Tin time after time, until the Calx has imbibed an eighth part of the Tutia, then reduce it into a Body, and you will find it to have Ignition, and that good: if not, reiterate the same labour, till due Ignition be acquired. All Waters dissolutive of Bodies and Spirits, we shall hereafter shew you every one of them according to their kind.

XIV. With Talck, or Mercury, or pure Luna (which is more profitable) deduced to this by calcining and dissolving, you may acquire the compleat Ignition and hardness of Saturn and Jupiter, with incomparable brightness; but Speculations in these things without practise, is not very available.

XV. To Grind, to Decoct, to Inhumate, to Calcine, to Fust, to Destroy, to Restore or Reduce, and to cleanse Bodies, are effectuall works: with these Keys you may open the Occult Inclosures of our Arcanum, and without them, you shall never sit down at the Repasts of satisfactions.

XVI. A White Medicine for *Jupiter and Saturn* prepated. *Take of fine Luna one Pound, living Mercury eight Pound, Amalgamate, and wash the Amalgama with spirit of Vinegar and common Salt prepared, until it acquires a Coelestial or Azure Colour. Then extract as much of Mercury as you can, by strongly expressing the mixture through a thick Cloth. To this add Mercury sublimate, double the wieght of the Luna, grind them well together, then decoct the mixture in a Bolt-Head, firmly closed for 24 hours: Decoct the same again, then break the Vessel, and then separate that which is Sublimed from the Inferior Redish Pouder. But take heed of giving too great a Fire, for that would cause the whole to flow into one black Mass. Put the Pouder upon a Porphyrie stone, add to it two parts of Sal Armoniack prepared, and one part of Mercury sublimed; grind all very well together, and imbibe the mixture* with the Water of Sal Alcali or Sal Nitre, if you cannot get the other, or Salt of Pot-Ashes: when imbibed, Distil off with a gentle Fire the whole Water, till that remains in the bottom is melted like Pitch: Cohobate the same Water, repeating this Work thrice. Then take out the Matter, grind it on a stone, and dry it very well: Imbibe again with rectified Oyl of Eggs, or with Sal Alkali, or Oyl or Salt of Pot-Ashes, or of Nitre, or Tartar, until it will flow with Ingress. Project one part upon five parts of Tin prepared, and it will be perfect Luna of the second Order, without Error.

XVII. Another White Medicine for *Jupiter and Saturn* prepared. *Take Talk Calcined, and grind it with as much as itself of Sal Armoniack; sublime it three or four times; dissolve into Water, and therewith Imbibe: Luna calcined (as you did in the former) so often as until it has drunk in as much as its own weight is, and give ingress to it with the Oyls aforesaid, and project one part upon 10 parts of Jupiter prepared, and it will be all fine Luna.*

XVIII. Another White Medicine for *Saturn and Jupiter* prepared. *Take Luna 1 pound dissolved in its own water (made of Nitre and Vitriol) to which add Taick calcined and dissolved 1 pound:*

Distil off the Water, cohobating 3 or 4 times, congeal and incerate with Arsenick sublimed, until it flow and have Ingress: project 1 part upon 8 parts of Jupiter prepared, and 'twill be all fine Luna. These three Medicines you may project upon Saturn prepared for the White, but then the Saturn must be prepared for the White, but then the Saturn must be prepared and calcined for three days, by Sect. 21 of the former Chapter.

XIX. A Solar Medicine for *Jupiter and Saturn prepared. Calcine Sol, amalgating first with Mercury, as in* Luna, *express the Mercury through a Cloth, then grind it with twice so much as itself of common Salt prepared; set the whole over a gentle fire, that the remaining Mercury may receed. Extract the Salt with sweet water, dry the Calx, from which sublime as much Sal Armoniack, reverting the sublimed Salt four times; dissolve it in A. F. made of Vitriol, Nitre, and Alum; dissolve also Crocus Martis made by calcination, or Copper Calcined red: joyn these Waters in equal parts; draw off the Water by distillation, and cohobate four times: then dry the matter and imbibe it with Oyl of Tartar rectified (as heretofore is taught) until it flows as Wax, and by projection will tinge four parts of Saturn or Jupiter into Gold Obrizon.*

XX. Another Solar Medicine for Saturn and Jupiter prepared for the Red. It is made of Sol *dissolved, Sulphur dissolved, and Verdigrise dissolved, mixt and prepared (as in the last Sect). and then incerated with Oyl of Hair prepared; or of Eggs, (for both are one) one part of this projects upon 10 parts of Saturn or Jupiter prepared for the red, and it will be most fine Gold according to its degree, these Medicines only altering in the second Order.*

XXII. There is also another preparation of Jupiter by Sect. 22. of the former Chapter.

XXIII. And in Sect. 23. of the former Chapter, you have the Regimen of Jupiter for the White, which generates or produces fine Luna, such as being tryed upon the Test, produces a Body perfect in Whiteness, and perpetually generating its life.

CHAPTER XLIV

OF THE ALCHEMY OF MARS.

I. To *prepare* MARS or Iron. *Calcine it as Venus with common Salt cleansed, and let it be washed with pure Vinegar; Being washed, dry it in the Sun, and when dried, grind and imbibe it with new Salt and Vinegar, and put it into the same Fornace* (as we shall direct in Venus) for 3 days. Esteem and value this Solution, viz. The water of fixed Sulphur, wonderfully augmenting the color of the Elixir.

II. The whole Secret of Mars is from the Work of Nature, because it is a Metalick body, very livid, a little Red, partaking of Whiteness, not pure, sustaining Ignition, fusible with violent fire extensive under the Hammer, and sounding much.

III. It is hard to be managed by reason of its impotency of fusion; which if it be made to flow by a Medicine changing its nature, is so conjoyned to Sol and Luna, that it cannot be separated by examen without great Industry: but if prepared, it is conjoyned, and cannot be separated by any Artifice, if the nature of that fixation be not changed by it, the defilement of the Mars being only removed. Therefore it is easily a Tincture of Redness; but difficulty of

Whiteness. And when it is conjoyned, it is not altred, nor does it change the colour of the commixtion, but augments it in quantity.

IV. Among all Bodies Jupiter is more splendidly, more clearly, more brightly, and more perfectly transmuted into a Solar or Lunar Body, than other Bodies, but the Work is of long labour, though easie to be handled: Next to Jupiter is Venus chosen, of more difficult handling, but of shorter labour than Jupiter. Next after Venus comes Saturn, which has a diminished perfection in Transmutation, and is easie to be handled, but of most tedious labour. Lastly, Mars among all the Bodies of least perfection, is in transmutation, most difficult to be handled, and of exceeding long labour.

V. And the more difficult any Bodies are of fusion, the more difficult they are in handling in the Work of Transmutation; the easier to be fused, the easier to be handled: and what diversity of perfections are found in any particular, in the lesser, or middle Works; yet in the Great Work all Bodies are of one perfection, but not all of a like easie handling or labour.

VI. Hence it appears, that Mars or Iron, is a commixture of fixed Earthy Sulphur, with fixed earthy

Argent Vive of a livid whiteness, the highly fixed Sulphur predominating, which prohibits fusion: Whence it is evident, that fixed Sulphur hinders fusion more than fixed Argent Vive: But Sulphur not fixed, hastens fusion more than unfixed Argent Vive: By which the cause of speedy or flow fusion in everybody is seen.

VII. What has more of a fixed Sulphur is harder to fuse, than what partakes of a burning fugitive Sulphur; which appears because Sulphur cannot be fixed without Calcination, and no Calx gives fusion, therefore in all things it, viz. fixt Sulphur, must impede the same.

VIII. The causes of the corruption of the Metals by fire, are:

1. The inclusion of a burning Sulphur in the profundity of their substance, diminishing them by Inflammation, and exterminating into Fume, whatever fixed Argent Vive was in them.
2. A Vehemency of the Exterior flame, penetrating, and resolving them, with itself into Fume, and the most fixed matter in them.
3. The rarefaction of them by calcination, the flame or fire, penetrating into, and exterminating them. Where all these causes of Corruption concur, those Bodies must be exceedingly corrupted. Where they all

concur not, they are by so much the less corrupted.

IX. The causes of the goodness of Bodies, is their abounding with Argent Vive. For seeing Argent Vive, for no cause of Extermination, will be divided into parts in its composition (because it either with its whole substance flies from the fire, or with its whole substance remains permanent in it;) it is necessarily concluded to be a cause of Perfection.

X. Therefore Praised and Blessed be the most Glorious and High God, who created it, and gave it a Substance and Properties, which nothing else in the World does possess besides; that this perfection might be found in it, (by the help of Art) as we have found therein with great power. For it is that which overcomes Fire, and by Fire cannot be overcome, but in it amicably rests, and rejoyces therein.

XI. Mars is prepared either with sublimation, or without sublimation, with sublimation we endeavour to unit it with Arsenick not fixed, as profoundly as we can, that in fusion it may melt with the same; but afterwards it is sublimed in a proper Vessel of sublimation, the which is the best and most perfect of all other Preparations. Mars is also prepared by Arsenick oftentimes sublimed from it, until some quantity of the Arsenick itself remain: For if this be reduced, it will flow out white, clean, fusible, and well prepared: Mars is also prepared by fusion of it with Lead and Tutia, for from these it flows clean and white.

XII. To Indurate or harden soft Bodies. Argent Vive precipitated must be dissolved, and the calcin'd Body (which you have a design to harden) dissolved likewise: mix both these solutions together, and the calcin'd body mixed with them by frequent imbibitions, etc. continually grinding, imbibing, calcining and reducing, until it be made hard and fusible with Ignition. The very same may also be compleatly effected, with the Calx of Bodies, and Tutia, and Marchasite, calcined, dissolved, and imbibed. The more clean these are, the more perfectly do they change.

XIII. To soften hard Bodies, as Mars, etc. They must be conjoyned and sublimed often with Arsenick, and after sublimation of the Arsenick, assated, or calcined with their due proportion of fire, the measure of which we shall declare in our Discourse of Fornaces. Lastly, They must be reduced with the force of their proper fire, until in fusion they grow soft, according to the degree of the hardness of their Bodies. All these alterations are of the first Order, without which our Magistery is not perfected.

XIV. Medicines dealbating Mars, of the first Order. That which dealbates it, of the First Order, is that which makes it to flow: The special fusive of it is Arsenick of every kind: But with whatsoever it is dealbated and fused, it is necessary it be conjoyned and washed with Argent Vive, until all its impurity be removed, and it be white and fusible. Or else let it be red hot with vehement ignition, and upon it Arsenick projected; and when it shall be in flux, cast a quantity of Luna thereon; for when that is united with it, it is not separated therefrom, by any easie Artifice.

XV. Or thus: Calcine Mars, and wash away from it all its soluble Aluminosity (inferring corruption) by the way of solution, but now mentioned (with Argent Vive) then set cleansed Arsenick be sublimed from it, and reiterate that sublimation many times, until some part of the Arsenick be fixed therewith. Then with a solution of Litharge mix, imbibe, grind, and moderately calcine, several times: And lastly, reduce it with the Fire we mentioned in the Reduction of Jupiter from its Calx; so will it come forth white, clean and fusible.

XVI. Or, Only with sublimed Arsenick, in its Calx, let it be reduced, and it will flow out white, clean and fusible: But here observe the Caution we shall give in the Chapter of Venus, concerning the reiteration of the sublimation of Arsenick, (fixing itself in its profundity) from it. Mars is likewise whitened after the same manner with Marchasite and Tutia.

XVII. To prepare Mars. *Grind one pound of the filings thereof, with half a pound of Arsenick sublimed; imbibe the mixture with the water of Salt Peter and Sal Alcali, reiterating this Imbibition thrice, then make it flow with a violent fire, so will it be white: Repeat this so long till it flow sufficiently, with a good whiteness.*

XVIII. The first White Medicine for Mars and Venus. Take *Silver calcined 1 pound, Arsenick prepared 1 pound, Mercury precipitation 1 pound, grind them together, and imbibe the whole with water of Salt Nitre, Litharge, and Sal Armoniack, in equal parts, (I suppose there is mean Aqua Regis) till it has drunk in its own weight of that water: Then dry, and incerate with white Oyl (as in others) until it flow, and one part full upon 4 parts of Mars or Venus prepared.*

XIX. The second White Medicine for Mars and Venus. *Take Luna calcined, Jupiter calcined and dissolved, ana: mix, dry, and increase with double their quantity of Arsenick sublimed, until the Medicine flows well.*

XX. The third White Medicine for Mars and Venus. Take Luna calcined, Arsenick and Sulphur sublimed, and ground with it, and then sublimed with a like quantity of Sal Armoniack. This sublimation repeat thrice, and then project 1 pound upon 4 pound of Mars or Venus prepared.

XXI. A Red, or Solar Medicine for Mars and Venus. *Take Tutia 1 pound, Calcine or dissolve it in AF, then with that water imbibe the Calx of Sol, that it may drink in double its own weight of the same water: Afterwards by distillation draw off the same water from it, cohabating four times. Lastly, incerate with Oyl of Han, or Bulls Gall, and Verdigrise prepared, and it will be excellent.* But be sure to pursue the Operation according to our Directions, otherwise you will labour in vain, and in your heart understand our Intentions (expressed in our Volumes) so will you know truth from falsehood.

XXII. To Calcine Mars. Mars being filed, is calcined in our Calcinatory Fornace, until it is very well rubified, and becomes a pouder impalpable without grinding. And this is called Crocus Martis.

XXIII. The Regimen of Mars. Take *of the Paste of Mars 2 pound, of the Pasts of Venus and Saturn, ana 3 pound, mix these without Ferment, and decoct the mixture for seven days, and you will find the whole dry. Fix it and add to it half its weight of Litharge in powder, which put into a Reductory Fornace, so will you have a Mineral substance very profitable if you be wise.*

CHAPTER XLV
OF THE ALCHYMIE OF VENUS.

I. The *Preparation of* VENUS. *Lay thin Copper Plates stratum superstratum with Common Salts prepared, till the Vessel be full, which cover, firmly Lute, and calcine in a fit Fornace for 24 hours:* Then take it out, scrape off what is calcined, and repeat the calcination of the Plates with new Salt as before, repeating the Calcination so often till all the Plates are consumed. For the Salt corrodes the superfluous humidity, and combustible sulphureity; and the fire elevates the fugitive and inflamable substance with due proportion. *This Calx grind to a most subtil pouder, wash it with Vinegar, till water will come from it free from blackness. Again, imbibe it with more Salt and Vinegar, and grind, and then calcine again in an open Vessel for 3 days and nights:* Take it out, grind it subtily and long, and wash it with Vinegar, till it is cleansed from all uncleanness. *This done, dry it in the Sun: Add to it half its weight of Sal Armoniack, grinding it long, to an impalpable substance: Then expose it to the Air, or set it in Horse—dung to be dissolved: To what is undissolved add a new clean Sal Armoniack; thus continuing till the whole be made water.* Esteem and value this water, which we call the water of fixed

Sulphur, with which the Elixir is tinged to infinity.

II. Venus is a Metalick Body, livid, pertaking of a dusky redness, subject to ignition, fusible, extensible under the Hammer, but refusing the Cupel and Cement. It is in the profundity of its substance of the color and essence of Gold, and is hammered being red hot, as Silver and Gold is. It is the medium of Sol and Luna, and easily converts it nature to either, being of good conversion, and of little labour.

III. It agrees very well with Tutia, which citrinizes it with a good yellow, from whence you may reap profit: we need not labour to indurate it, or make it ignitible, therefore it is to be chosen before other imperfect Bodies, in the lesser and middle Work, but not in the greater. Yet this has a Vice beyond Jupiter, that it easily grows livid, and receives foulness from sharp things, to erradicate which, is not an easie, but a profound Art.

IV. Copper therefore is unclean Argent Vive, mixed with Sulphur unclean, gross, and fixed, as to its greater part; but as to its lesser part, not fixed, red, and livid, in relation to the whole, not overcoming nor overcome. Its volatile Sulphur is evident from its sulphurous fume, and loss of quantity by frequent fluxing and combustion. Its fixt Sulphur is evident from its slowness of fusion, and induration of its substance. And that there is an unclean red Sulphur joyned with unclean Argent Vive, is evident even to the senses.

V. When the fixed Sulphur comes to fixation by heat of Fire, its parts are subtilized; but that part which is in the aptitude of solution of its substance is dissolved; the sign of which is the exposing it to the vapours of Vinegar, which makes the Aluminosity of its Sulphur flow in its Superficies. And being put into a saline liquor, many parts of it are easily dissolved by Evulition; this Aluminosity by a saline watriness, and easie solution, is changed into water: For nothing is watery, and easily soluble, except Alum, and what is of its nature. This understand also of the body of Iron.

VI. But the blackness in either Venus or Mars, created by the Fire, is by reason of the Sulphur not fixed, (much indeed in Venus, but little in Mars) and it approaches nigh to the nature of fixed Sulphur. Hence it is evident, that fusion is helped, and partly made by Sulphur not fixed, but hindred from Sulphur fixed. This he certainly knew to be true, who by no art of fusion could make Sulphur to flow after its fixation:

71

But having fixed Argent Vive, by frequently repeating the sublimation thereof, found it apt to admit good fusion.

VII. Hence it is evident that those Bodies are of greater perfection, which contain more of Argent Vive, those of lesser perfection which contain lesser. Therefore study in all your Works to make Argent Vive to exceed in the Commixtion. And if you could perfect by Argent Vive only, you would have attained to the highest perfection, even the perfection of that which overcomes the Works of Nature: For you may cleanse it more inwardly, to which purification nature cannot reach.

VIII. This is manifest; for that those Bodies which contain a greater quantity of Argent Vive, should be of greater perfection, arises from their easie reception of Argent Vive into their substance: and we see Bodies of perfection amicably to embrace each other.

IX. Out of what has been said it is also apparent, that in Bodies there is a two-fold sulphureity: One indeed included in the profundity of Argent Vive, in the begining of their mixtion: The other supervenient from other Accidents. The one of them may be removed with labour; but the other cannot possibly be taken away by any Artifice or Operation of the Fire, to which we can profitably come, it being so firmly and radically united therein. And this is proved by experiment; for we see the aductible sulphureity to be abolished or destroyed by fire, but the fixed sulphureity not so.

X. Therefore when we say, Bodies are cleansed by Calcination, understand that to be meant of the earthy substance, which is not united to the Radix of their nature: For it is not possible by Art, or force of fire, to cleanse or separate what is united, unless the Medicine of Argent Vive has access.

XI. Now the separation of an earthy substance from its compound, which in the root of nature is united to a Metal, is this: Either it is made by elevation, with things elevating the substance of Argent Vive, and leaving the sulphureity, by reason of its conveniency, with them: of which nature are Tutia and Marchasite; because they are Fumes, part of which has a greater quantity of Argent Vive than of Sulphur.

XII. The proof of this you may see, when you joyn those things with Bodies in a strong and sudden fusion, for these Spirits in their flight, carry up the Bodies with them; and therefore you may elevate them with them. Or else, by a Lavation or Commixtion with Argent Vive, as we have already said: For Argent Vive holds what is of its own nature, but casts out what is alien or foreign.

XIII. The preparation of Venus. It is manifold; one by Elevation, another without Elevation. The way by Elevation is, that Tutia be taken (with which Venus well agrees) and that it be ingeniously united therewith: Then put it into a Vessel of sublimation to be sublimed; and by a most exceeding degree of Fire, its most subtil part will be elevated, which will be of most bright splendor. Or, it may be mixed with Sulphur, and then elevated by sublimation.

XIV. But without sublimation, it is prepared either by cleansing things in its Calx, or in its Body: As by Tutia, Salts, and Alums: Or, by a Lavament of Argent Vive, as all other imperfect Bodies are.

XV. *The Preparation, or Purgation of Venus, also is two—fold, viz, one for the White, and the other for the red; for the White it is thus. Take Venus calcin'd by fire only (as aforesaid) ground fine 1 pound: Arsenick sublimed 4 ounces: Grind them together, and imbibe the mixture 3 or 4 times with water of Litharge, and reduce the whole with Sal Nitre, and Oyl of Tartar, and you will find the Body of Venus white and splendid, and fit for receiving its Medicine.*

XVI. The Preparation for the Red. *Take filings of Venus 1 pound, Sulphur 4 ounces, grind them together: Or cement Plates of Copper with Sulphur, and so calcine: wash the calcin'd with water of Salt and Alum; and then with things reducing, reduce it into a body, clean and fit for the reception of the Red Tincture.*

XVII. Another Preparation for the Red. *Calcine it with fire only, and then dissolve a part thereof, and likewise dissolve a part of Tutia calcin'd; joyn both solutions, and with the same imbibe the remaining part of the Calx of Venus 4 or 5 times: Or, you may make this Imbibition with Tutia alone dissolved, provided that more of the Tutia (than half of the Calx is) be imbibed in the said Calx. This done, reduce with things reducing, and you will have the Body of Venus clean and splendid;* which with a little help may be brought to an higher state, if you have studiously penetrated into the Truth.

XVII. Another Preparation for the Red. *of Venus calcin'd per Se, or with the fire alone you may make an intense greatness, called Flos Cupri vel Veneris: Dissolve this greenness in Spirit of Vinegar, and then congeal it; afterwards with things reducing, reduce the congelate, which when reduced, will be a Body fit for many Works.*

XIX. *Medicines dealbating Venus, of the first Order.*
There is one Medicine for Bodies, and another for
Argent Vive, and of Bodies; one is of the first Order;
another of the second; and another of the third: and
so likewise the first, second, and third, of Argent
Vive. Of the Medicine of Bodies of the first Order, we
say there is one of hard Bodies, and one of soft: of
hard Bodies, there is one of Mars (of which in the
former Chapter) one for Venus, of which in this place;
and one for Luna (of which in the next Chapter.) Of
soft Bodies, there is one for Saturn, and another for
Jupiter. That of Venus and Mars, is the pure
dealbation of their substance; but that of Luna the
rubification of it, with citrinity of a pleasing
brightness, which rubification is not given to Mars
and Venus, by Medicines of the first Order: For being
totally unclean, they are unapt to receive the
splendor of redness, before they are fitted with a
preparation inducing brightness. There is one Medicine
whitening Venus by Argent Vive, and another by
Arsenick. The Medicine of Argent Vive is thus made.
*Pirst, Argent Vive precipitated, is dissolved; then
calcined Venus dissolved likewise: These solutions are
mixed and after they are coagulated, they are
projected upon the Body of Venus.*

XX. Another way by Argent Vive. Argent Vive and
Litharge are dissolved a part, and the solutions

joyned together. Calx of Venus also is dissolved, and that solution joyned with the former, and then coagulated together, which projected upon Venus whitens it. Or thus. A quantity of Argent Vive is sublimed often from its body, till part thereof remain with it, with compleat ignition: and this mixture is very often imbibed and ground with Spirit of Vinegar, that it may the better be mixed in the profundity thereof, then it is assated, or moderately calcined, and lastly fresh Argent Vive is in like manner sublimed from it, and the remaining matter again imbibed, and moderately calcined as before, which work is so often to be repeated, till a large quantity of Argent Vive reside in it, with compleat ignition. This is a good dealbation of the first Order.

XXI. Another way thus. Argent Vive in its proper nature is so often sublimed from Argent Vive precipitated, till in it, the same is fixed, and admits good fusion: This fused matter projected upon the Body of Venus peculiarly whitens it. Or thus. A Solution of Luna, mixt with a solution of Litharge, coagulated, may be projected upon Venus; but is indeed better whitened if Argent Vive be perpetrated in all the Medicines.

XXII. The whitening of Venus with Arsenick of the first Order.

Take Calx of Venus, from it sublime Arsenick by many Repetitions, till it remains therewith and whitens it; but if you be not well skilled in the ways of sublimation, the Arsenick will not perservere in it without alteration: Therefore, after the first degree of sublimation, repeat the work in the same manner as in the sublimation of Marchasite. Chap. 40, Sect. 2.10. Or thus. Project Arsenick sublimed upon Luna, and then the whole upon Venus, it dealbates it peculiarly. Or, first mix Litharge, or burnt Lead, dissolved with Luna, and cast these upon Arsenick, and project the whole upon Venus, so will it be whitened; and this is a good dealbation of the first Order.

XXIII. Another way thus. *Upon Litharge alone dissolved and reduced, project Arsenick sublimed, and the whole upon Venus in flux, it whitens the same admirably, Or thus. Let Venus and Luna be commixed, and upon them project any of the above described dealbative Medicines.* For Luna is more friendly to Arsenick, than to any of the other Bodies, and therefore takes away fraction from it; and Saturn secondarily, and therefore we mix it with them. Also we melt Arsenick sublimed, that it may be all in a Lump, which being broken, we project piece after piece upon Venus: We do it in pieces, rather than in pouder, because the pouder is more easily inflamed, than a Lump, and so more easily Vanishes, before it can fall fiery hot upon the body.

XIV. In like manner, the Redness is taken away from Venus, and it is whitened with Tutia: But Titia suffices not, because it gives only a Citrine colour; which is yet of affinity to Whiteness. Any kind of Tutia is calcined and dissolved; and the Calx of Venus, also: These Solutions are conjoyned, and with them, the Body of Venus is citrinated. If you be well skill'd in this Work, you will find profit. Or thus. Take Marchasite sublimed, and proceed with it as with Argent Vive sublimed; the way is the same, and it whitens well.

XXV. To make the White and the Red Medicines for Venus. They are exactly made by the Rules of Prescripts delivered in Chap. 44 Sect. 19, 20, 21, 22. aforegoing, to which, I shall here refer you; for the Operations of those Medicines both for the White and Red, in the Bodies of both Mars and Venus, are one and the same.

XXVI. To Calcine Venus. *Take Filings of Copper, and put them to calcine either per se, or with Arsenick poudred, or with Sulphur, being anointed with common Oyl, calcine 3 or 4 days with a most strong fire: strike what is calcin'd that it may fall from the*

*Plates, (if you use Plates) which again calcine. The
Calx beat fine, re-calcine it, till it is well
rubified, and keep it for use.*

XXVII. The Regiment of *Venus and Saturn. Take of the
Paste of Venus, 3 Pounds; of Saturn, 2 Pounds; of the
Ferment, 1 Pound: of these, perfectly dissolved, make
a commixtion through their least parts, which keep in
sufficient heat, as in the White is said. Extract the
Water, and what remains in the Cloth, put into a well
sealed Glass, for 3 Weeks: Then take it out, and add
to it a third part of its own reserved water, and
decoct by Chap. 42 Sect. 23. aforegoing', which Work
do thrice. When it has imbibed all its proper Water,
put it in Its proper Vessel and Fornace to be fixed.
When fixed, with things, reducing, reduce it into a
Body, ready to be reduced and tinged.*

XXVIII. We more especially handling the Regimen of
Venus, do declare, that you ought seven times, or
oftner to rectifie it, when prepared and dissolved,
distilling off the Water, and cohobating thereon each
time, which being coagulate, thence make a most noble
Greenness, with Sal Armoniack dissolved in Spirit of
Vinegar. That greenness rubifie in a Vessel of Mars,
and again dissolve it, to which solution adjoyn a
third part of prepared and dissolved Luna; afterwards
extracting and cohobating the water of Ferment 7

times. Then reduce this into a Body, and you will rejoyce. The Regimen of Mars, is as of Venus, but by reason of its foulness, no great good arises from it.

XXIX. *Grind Luna, amalgamated with Mercury, with twice so much Metaline Arsenick (Quaere, Whether Regulus of Arsenick be not intended?) To which adjoyn a tenfold proportion of Venus amalgamated with Mercury: Grind the whole, and fix, and reduce into a Body, so will you have a pure White Metal.*

XXX. The first Dealbation of *Venus. Take Realg-ar 1 ounce, Argent Vive sublimed 3 Ounces and half, Tartar calcin'd, 1 ounce, grind and incorporate, put them into a Bolt head, a Foot and half high, and its Orifice so wide, as two Fingers may go into it: lute it, and set it over a Fire, covered with a Cloth: First make a gentle Fire for a quarter of an hour, afterwards augment the Fire underneath, and round about until the Fornace be very hot with Ignition; when all is cold, break the Vessel, and take out what you find Metalline;* and make of this a great quantity.

XXXI. A Second Dealbation, *Upon Tutia sublime one part of Mercury sublimate, and two parts of Arsenick sublimed, until it shall have ingress.* This clearly and very speciously whitens Venus.

XXXII. A Third Dealbation, Take Mercury *sublimate 3 Ounces, Arsenick sublimed 2 Ounces, dissolved with Litharge, till they become 8 Ounces: to these 8 Ounces, adjoyn other 8 Ounces of Arsenick sublimed, grind them together, and flux them with Oyl of Tartar, and therewith you may whiten prepared Venus at pleasure.*

XXXIII. A Fourth Dealbation, *Grind Metaline Arsenick, with as much of the Calx of Luna, and imbibe the Mixture with the Water of Sal Armoniack, and dry and grind: then dissolve Salt of Tartar, in the Water of Salt Nitre (some suppose Spirit of Nitre) with which Oyl imbibe the Medicine: repeat this thrice, incerating and drying, and you will rejoyce.*

XXXIV. A Fifth Dealbation, which is of our own Invention.

Imbibe Jupiter calcined, washed and dryed, so often with metal-me Arsenick, and half so much Mercury sublimate, as untill it flows and enters Venus, which, (if first prepared) it whitens speedily.

XXXV. A Sixth Dealbation upon *Tutia calcined, dissolved and Coagulated, sublime White Arsenick (so that the Arsenick be 3 parts to 1 of the Tutia) reiterating the sublimation upon it four times; for it* has Ingress. With them mix half as much as the whole is of Mercury sublimate; grinding and incerating 4 times with the Water of Sal Armoniack, Nitre and Tartar, ana. (Quere, whether that may not be Aq. Regis) with this when coagulated, cement prepared plates of Venus, and melt, so will you have a very beautiful Body.

XXXVI. A Seventh Dealbation. *Grind Venus, calcined and incerated, adding to it Arsenick sublimed, and half a part of Mercury sublimate; with which being well ground and mixed, add a little of the Water of Sal Armoniack (Quer. if not A.R.) incerating upon a marble; after dry and sublime. Revert the sublimate upon the Foeces, again imbibing, which do thrice: the fourth time imbibe with Water of Nitre (Spirit of Nitre) and sublime what can be sublimed: reiterate this Labour till it remains fluid in the bottom. This in Copper prepared, will be Resplendent with brightness.*

XXXVII. An Eighth Dealbation. *Upon the prepared Calx of Venus, so often sublime Arsenick sublimate, till some part of the Arsenick remaine with it in the strongest Fire. That imbibed with the Water of Nitre (Spirit of Nitre) and lastly incerated with Water of Luna, and Mercury precipitate, and in the end with Oyl of Tartar Rectified, until it flows, wonderfully whitens Venus, and enters the second order, if you have operated right. For I have elsewhere said, that if you obtain any part of Mercury precipitated, in the mixture, your Work will be more splendid: especially, if the White Ferment, dissolved with the Mercury dissolved, after a certain fixation of it, be added by the medium of Inceration; by which you will find you have traced the high way itself.*

Geber, the Author, here saith, that the last 8 Sections are all proved Experiments: the first 4 of them, being Experiments of the Ancients, by him again proved; the latter 4, Rectifications of the Practices of the Ancients, or rather Experiments of his Own: All which he affirms to be absolutely true, and by him proved so.

CHAPTER XLVI
OF THE ALCHYMIE OF LUNA.

I. The *Preparation of* LUNA. It is subtilized, attenuated and reduced to a Spirituality in the same manner, as hereafter in Chap. 47 Sect. 1 we shall teach concerning Sol. Therefore in all, and every part of the Work, do the same as we shall there teach with Gold: and this work of Luna dissolved, is the *Ferment* for the White Elixir made Spiritual.

II. It is a metalick Body, white, which pure whiteness clean, hard, founding, very durable in the Cupel, extensible under the Hammer, and fusible. It is the Tincture of whiteness, hardens Tin by Artifice, and converts it to itself; and being mixed with Sol, it breaks not, but in the examination, it perseveres without Artifice.

III. He who knows how to subtilize it, and then to inspissate and fix it associated with Gold, brings it into such a State, that it will remain with *301* in the Test, and be in no wise separated from it, being put over the fumes of sharp things, as Vinegar A. F. or Salarmoniack, and it will be of a wonderful Caelestine Color: It is a noble Body, but wants of the Nobility of Sol, and its *Minera* is found determinate; but it has often a Minera confused with other Bodies, which Silver is not so Noble. It is likewise dissolved and Calcin'd with great Labour, and no Profit.

IV. If therefore clean, fixed, Red and clear Sulphur, fall upon the pure substance of Argent Vive, thereof is made pure Gold; then in like manner, if clean, fixed, white and clear Sulphur, falls upon the substance of Argent Vive, there is made pure Silver, if in quantity it exceed not: yet this has a purity short of the purity of Gold, and a more gross inspissation than Gold hath; the sign of which is, that its parts are not so condensed, as that it can be equal in Weight with Gold, nor has it so fixed a substance as that; which is known by its diminution in the Fire; and the Sulphur of it, which is neither fixed nor incombustible, is the cause of that diminution.

V. But it is not impossible or improbable to give Judgment of the same, as fixed and not fixed, in the respect of one Body to another: for the Sulphur of Luna compared with the Sulphur of Sol, is not fixed and burning; but in respect of the Sulphur of other bodies, it is fixed and not burning.

VI. The Citrinating of Luna, by medicines of the first Order:

This is that which adheres to it in its profondity, and adding color either by its proper Nature, or by the Artifice of this Magistery. We declare therefore that Medicine which arising from its own root, adhers to it; but there are Artifices by which we make a thing of every kind to adhere with firm ingress. But Our Medicine we extract either from Sulphur, or Argent Vive, or a commixture of both: from Sulphur less perfectly; but from Argent Vive more perfectly. This Medicine may also be made of certain mineral things, which are not of this kind; as of Vitriol, and Copperas, (which is called the Gum of Copper.)

VII. The method by Argent Vive. Take Argent Vive precipitated, viz, mortified and fixed by precipitation, put it into a Fornace of *great Ignition, (after the manner of Conservation of Calces) until it be red as Usifur, (Cinabar.) But if it be not red, take a part of Argent Vive not mortified, and with Sulphur reiterate the sublimation thereof: The Sulphur and Argent Vive must be cleansed from all impurity: Repeat the sublimation of it twenty times upon the praecipitate, then dissolve it with dissolving water, and again calcine and dissolve, till it be Exuberally done. Then dissolve a part of Luna, mix the Solutions, and coagulate them, and project the coagulated matter upon Luna in flux, and it will colour it with a peculiar Citrinity. But if Argent Vive be in its precipitation Red, the aforesaid Administration, without commixtion of anything tinging it, is sufficient for the compleatment of its perfection.*

VIII. The Method by Sulphur, is difficult, and irnmensly laborious. It is Citrinated with a solution of Mars, but then you must first calcine it, and then fix it with abundance of Labour, then administer it with the same preparation, and the same projection upon the Body of Luna: But hence results not a splendid bright colour, but a dull, and livid, with a mortiferous Citrinity.

IX. The Citrinating of it with *Vitriol, or Copperas. Take of either of them, q.v. and sublime as much thereof as can be sublimed, until the fire be increased to the highest degree. Then sublime this sublimate, with a fit fire, that of it, part after part may be fixed, until its greater part be fixed. Afterwards warily calcine it, that a greater fire may be administred for its perfection: This done, dissolve it into a most red Water, (which has no equal) and so operate, that you may give it ingress into the Body of* Luna. These three last Sections, are all Medicines of the first Order.

X. We thus seeing things of this kind, profoundly, and amicably to adhere to Luna, have considered, (and it is certain) that these are from its own Radix; and thence it is, that Luna is altered by them. It is also to be noted, that Medicines of Argent Vive, if they alter Luna with more than one only difference, in order to a total Compleatment: They are not of the first Order.

XI. *A Lunar Medicine of the third Order for the White.* It is as well for perfecting imperfect Bodies, as for coagulating Mercury itself into true Luna: And is thus made. Take *Luna calcined, dissolve it in solutive water (Aqua fortis,) then decoct it in a Phial with a long Neck, the Orifice of which must be left unstopt, for one day only; until a third part of the water be consumed: Then put the vessel into a cold place, to convert into fusible Crystals,* or Vitriol. This is Silver reduced to our Mercury, fixed, and fusible. Take *of this 4 Ounces, of White Arsenick prepared 6 Ounces, Sulphur prepared 2 Ounces; mix altogether well grinding them with Nitre and Sal Armoniack; put the mixture into a Bolt-heat, keeping the same in heat for a Week, that the matter may be hard as Pitch. This take out, and again incerate the third time, and in 3 days you will find it an Oyl in flux: when the vessel is cold, break it, and take what you find therein, which will be in a lump fixed, and flowing as Wax.*

This is the first degree. Again, Take of new Matter, as much as before; and joyn the same with this ferment, and do as before; and consequently, a third, and a fourth time. Thus doing, you will find a Medicine, which is great and excellent in goodness; for 1 part falls upon 10 of any other Body, or of Mercury, and converts it into true Luna. Keep this Stone, and considerately luminate upon the things we teach, and you will attain unto higher things.

XII. A Lunar Medicine of the third Order for the White. Take *the known Stone of it, and by way of separation, divide its most pure substance and keep it apart. Then fix some of that part, which is most pure, leaving the remainder, and when it is fixed, dissolve what is soluble of it; but what is not soluble, put to be calcined, and again dissolve the calcinate, until again what is soluble of it be altogether dissolved. Continue this process until the greater quantity be dissolved. Then mix all the solutions together, and coagulate them; this done, gently decocting, keep the coagulate in a temperate fire, until greater fire may be fitly added for its perfection. Therefore reiterate all these Orders of Preparation upon it 4 times; and lastly, calcine it by its own way;* for thus administring you have sufficiently governed, the most precious Earth of the Stone.

Then subtily, and ingeniously conjoyn a quantity of the part reserved, with part of this prepared Earth, through its least Particles, then sublime by way of sublimation, until the fixed with the not fixed, be wholly, elevated; which if you see not, again add a quantity of the not fixed part, until enough be added for elevation thereof. When it is all sublimed, repeat the sublimation, until be repetition of this OperatiQn, it be wholly fixed. Being fixed, again imbibe it with quantity after quantity of the not fixed, after the same manner, till the whole shall be again sublimed, then again fix it, until it have easie fusion with Ignition. This is the true Medicine which transmutes all imperfect Metals, and every Argent Vive into most fine and perfect Luna.

XIII. The Regiment of Luna. Dissolve and Coagulate it 7 times, or at least 4 times; and to it dissolved, adjoyn the fixed Rubifying Waters, which we shall declare, and you will find the body aptly solar, for it agrees with Sol, and remains quietly with it. In this, Venus admirable well purged and dissolved, may be a great help to you, because a most clean, tinging, and fixed Sulphur may be extracted from it. And I tell you, that Mercury purified and fixed, has power to palliate, or illustrate the foulness of imperfect Bodies; and fixed Sulphur extracted pure from Bodies, to tinge them with splendor.

XIV. Hence you may gather a great Secret, viz. That Mercury and Sulphur may be extracted as well from imperfect Bodies, as from perfect: For purified Spirits, and middle Minerals are an help, and very peculiar for deducing the Work to perfection.

XV. *Another Regiment of Luna. This is to reduce it to a more noble state. Take Luna dissolved 3 Pounds, of Venus dissolved 4 Pounds, of Ferment dissQlved 1 Pound; conjoyn the dissolutions decoct them for 7 days, with gentle fire, in a sealed glass, as in Mars, with their whole water; then augment the fire leisurely for other 7 days, and let it be as a fire of Sublimation. For other 7 days give it fire yet stronger, that the whole water may be fixed with it. This pouder reduce in a small quantity; and if it retains with itself part of the Mercury, (which you will easily perceive if you know how to calcine). it is well indeed; but if not put it again to be fixed, until it is sufficiently fixed.* This must be reduced with red reducing Medicines, so will you find your Luna, tinged, transmuted, and fixed.

XVI. The Ferment of Luna for the White. It is made by dissolving Luna in its own Corrosive water, and then boiling this water away to a third part, it is to be exposed to the Air, or set in B. M. or in Dung for certain days; so will it be Oyl of Luna, and Ferment, which keep for the White Work.

XVII. The Ferment of Ferments upon *Mercury for the White. Take of the Ferment of Luna, which is its Oyl; add to it twice as much of Arsenick sublimed, and dissolved in water, (Quaer. what Water?) then to both these add of Mercury dissolved, as much as of the Arsenick: mix the Waters, set them over the fire for one day to be incorporated, then draw off the water by an Alembick, and cohobate fifteen times; so incerating, it will be fluid as fusible Wax. Add to it as much Virgin-Wax melted; commix them, and project the mixture upon Mercury washed, (Quaere, What is meant by washing here?) according as you see fit: for that resolved, is augmented in Virtue and Weight.*

XVIII. A Work upon *Luna and Mercury. Take Litharge, Salt of Pot—Ashes, mix and make a Cement: Put the Cement first into a Crucible an Inch thick, upon which put a Ball of the Amalgamation of Mercury and Luna; upon which, put the remainder of the Cement, that the Ball may be in the middle: Dry, lute, and set the Crucible in a gentle fire half a day, leisurely, augmenting the fire and so continue its leisurely increase, from the Evening unto the dawning of the day, with moderate Ignition at last; then take it out, and prove it by Cineritium, and it will be Luna in weight and surdity, and much better in fixation.*

XIX. Another Work. *Amalgamate Luna with Mercury, to which add as much Saturn, as there is Luna; put it into such a Crucible that a fourth part of it may be empty: Assuse on it Oyl of Sulphur, and decoct it unto the consumption of the Oyl: Afterwards keep it for two hours in a moderate fire; and there will be generated a black Stone, with a little Redness. This Stone prove by Cineritium., and you will find your Luna augmented in Weight, Surdity, and Fixation.*

XX. Another Work: Take Luna *amalgamated with Mercury. Grind it with twice so much Metaline Arsenick, to which a tenfold proportion of amalgamated Venus, (viz.* That the Amalgamation of Venus, may be 10 times as much as the whole Amalgama of Luna and Mercury mixed, with the duple quantity of *Arsenick) grind the whole and fix: Then reduce it into a Body, and you will find a good augmentation.*

XXI. Of the Citrination of Luna, or tinging its Body yellow.

Dissolve our Philosophick Zyniar, (which is Verdigrise) deduced from Venus prepared, in the water of the dissolution of Luna, (Aqua Fortis) to which adjoyn half so much as its self is of Mercury rubified by sublimation, and in some sort fixed, and dissolved; to these add, as much of Luna dissolved, as the Zyniar (Verdigrise) is; from which (fermented for one day) extract the water by distillation, and cohobate 10 times, then coagulate and reduce into a body, and you will find it a good Work.

XXII. Or *thus.Dissolve Zyniar 1 Ounce, and our Crocus prepared with Mercury, sublimate till it wax red 1 Ounce; add as much Sal Armoniack, and sublime it thrice from that Crocus, which dissolve: To which add of Luna dissolved 2 Ounces: Then do as in the former, incerating and reducing, and you will find satisfaction.*

XXIII. Or thus. *Take of Crocus and Zyniar dissolved ana; add as much Sol dissolved, incerate as before, then coagulate; to the coagulate add a fourth part of its weight, of the Oyl of Salt-peter; and project upon so much of Luna, and will be a Tincture of a Citrine aspect.*

XXIV. Or thus. *Make a Water of our Zyniar, and of our said Crocus, and imbibe the Calces of Sol and Luna, of each equal parts, therewith, until they have drunk in* their own weight of it; Then incerate with the Oyl of Sal Armoniack, and Nitre, and reduce the Mass into a Noble Body.

XXV. Or thus. *Sublime Sal Armoniack from our greenness, to which add Crocus and Zyniar; from which well commixed, sublime the Sal Armoniack, and repeat it twice or thrice. Then dissolve the whole, to which add a third part of Gold dissolved; incerate as before and congeal; then project upon Sol 1 ounce, Luna 2 Ounces, mixed together, and it will be good.*

CHAPTER XLVII
OF THE ALCHIMIE OF SOL.

I. Perfect Bodies (as Sol is) need no preparation, in relation to their farther perfection; but that they may be more subtilized and attenuated, we give you this Preparation. *Take Leaves of fine Sol, which lay stratum superstratum, with common Salt well prepared, in a Vessel of Calcination; Set it into a Fornace, and calcine well for 3 days, until the whole be subtily calcined: Then take it, grind it well, wash it with Vinegar (Quaer. Whether Spirit of Vinegar, or some other acid Spirit?) and dry it in the Sun: Then grind it well with half its weight of prepared or purified Sal Armoniack, and set it to be dissolved, until the whole (by help of the Common Salt, and Sal Armoniack) is reduced into a most clear water.* This is the pretious ferment for the Red Elixir, and the true Body made spiritual.

II. Gold is a metalick body, citrine, ponderous, mute, fulgid, equally digested in the Bowels of the Earth, and very long washed with mineral water; under the Hammer extensible, fusible, and sustaining the tryal of the Cupel and Cement.

III. From this definition you may conclude, That

nothing is true Gold, unless it has all the Causes and Differences of the definition of Gold: Yet whatever Metal is radically Citrine, and brings to equality, and cleanses, it makes Gold of it; from whence we discern, that Copper may be transmuted into Gold by Artifice. For we see in Copper Mines, a certain water, which flows out, and carries with it thin scales of Copper, which by a long continued course it washes and cleanses: But after such water ceases to flow, we find these thin scales, with the dry Sand, in 3 years time to be digested with the heat of the Sun; and among those Scales the purest Gold is found. Therefore we judge, that those Scales were cleansed by the help of the water, but equally digested by the heat of the Sun, in the dryness of the Sand, and so brought to perfection.

IV. Also Gold is of Metals the most pretious, and it is the Tincture of Redness, because it tinges and transforms every Body. It is calcined and dissolved without profit, and is a Medicine rejoycing, and conserving the Body in Youthfulness. It is most easily broken with Mercury, and by the Odour of Lead. There is not any Body that in Act more agrees with it in their substance than Luna and Jupiter; but in weight, deafness, and putrescibility, Saturn, and in colour Venus:

But indeed Venus in Potency is nearer Luna than either Jupiter, or Saturn, then Saturn, lastly Mars, Spirits are also commixed with it, (viz. Sol) and by it fixed, but not without great ingenuity and industry, which the floathful Artist shall never attain to the knowledge of.

V. Of the Nature of Sol. It is created of the most subtil substance of Argent Vive, and of most absolute fixedness; and of most small quantity of Sulphur, clean, and of pure redness, fixed, clear, and changed from its own nature, tinging that. And because there happens a diversity in colours of that Sulphur, the Citrinity or Yellowness of Gold, must needs have a like Density.

VI. That Gold is of the most subtil substance of Argent Vive, is most evident, because Argent Vive easily retains it; for Argent Vive retains not anything which is not of its own Nature. And that it has the clear, and clean substance of that, is manifest by its splendid and Radiant brightness, manifesting itself not only in the Day, but also in the Night. And that it has a fixed substance, void of all burning Sulphureity, is evident by every Operation in the Fire, for it is neither diminished, nor inflamed.

VII. And that it is tinging Sulphur is manifest, for being mixt with Argent Vive, it transforms the same into a Red color: And being sublimed with strong Ignition from Bodies, so that the substance of them ascends, with them it creates a most Yellow color; and that it is yellow, is evident even to the sence itself.

VIII. Therefore the most subtil substance of Argent Vive brought to Fixation, and the purity of the same, and the most subtil matter of Sulphur, fixed and not burning, is the whole Essential matter of Gold.

IX. But in it is found a greater quantity of Argent Vive than of Sulphur: Therefore Argent Vive has greater ingress into it. For this cause, whatsoever body you would alter, alter them according to this Exemplar, that you may deduce them to the equality thereof. For Gold having a subtil and fixt part, those parts would in its Creation be much condensed; and this was the cause of its great weight. Now by great decoction made by nature, a leisurely and gradual resolution of it was made, together with good inspissation, and its ultimate mixtion, that it might melt in the fire.

X. From what has been said, it is evident, that a large quantity of Argent Vive, is the cause of perfection; but much of Sulphur is the cause of Corruption. And uniformity of substance, which through the mixtion, is made by a natural decoction is cause of perfection; but diversity of substance is the cause of imperfection. Also Induration, and Inspissation, which is made by a long and temperate decoction, is a cause of perfection, but the contrary, of corruption and imperfection. Therefore if Sulphur shall not duly fall upon Argent Vive, diverse Corruptions must necessarily be inferred, according to the diversity of it, as if it be all, or part of it fixed, or not fixed; all, or part of it adustible, or not adustible; all clean, or half unclean, or it be much or little in quantity, exceeding, or being diminished in proportion, neither overcoming nor overcome, White or Red, or between both: From all which Diversities, divers Bodies were generated in Nature.

XI. A Solar Medicine of the Third Order. It is made by the Additament of Sulphur, not burning, by way of fixation, and calcination prudently and perfectly administred, and by manifold repetition of solution, until it be rendered clean: For by the perfect doing of these things, its cleansing by sublimation will be compleated, Thus. *Reiterate the sublimation of the not fixed part of the Stone, with this said Sulphur, conjoyning them according to Art, till they be first elevated together, and then fixed so, as to abide in the heat of the fire without ascension.* The oftner this Order of compleating the Exuberancy, shall be repeated, the more will the Exuberancy of this Medicine by multiplied, and the more its goodness augmented, and the augmentation of the perfection thereof highly multiplyed also.

XII. The whole compleatment of the Magistery is thus. By the way of sublimation, the Stone and its Additament may most perfectly be cleansed, and then by the Laws of Art, the fugitive must be fixed in them: And in this order is compleated the most pretious Arcanum, which is above every secret of the Sciences of this World, and a Treasure inestimable. Dispose yourself by exercise to it, with great industry and labor, and a continued Depth of Medition; for by these you will find it, and not otherwise. And indeed, in the preparation of the Stone, the reiteration of the Goodness of Administration upon this Medicine, may with industrious wariness be so far available, as to enable it to change Argent Vive into an infinite true Solisick, and Lunisick, without the help of anything more than its Multiplication.

XIII. The most high Gold the maker of all things, blessed and Glorious, be praised; who has revealed to us the series and order of all Medicines, with the Experience of them, which through his goodness, and our incessant Labor, we have searched out; which we have seen with our Eyes, and handled with our Hands, even the whole compleatment of the Magistery. But if we have concealed anything, ye Sons of Learning wonder not; for we have not concealed it from you, but have delivered in such Language, as that it may be hid from evil Men, and that the unjust and Vile might not know it. But ye Sons of Doctrine, search, and you shall find this most excellent gift of God; which he has reserved for you. Ye Sons of folly, impiety and prophaneness, avoid you the seeking after this Knowledge, it will be Enimical and destructive to you, and precipitate you into the State of Contempt and Misery. This gift of God is absolutely, by the Judgment of the Divine providence, hid from you and denyed you forever.

XIV. A solar Medicine of the third Order. It is made of Sol dissolved and prepared after the manner of Luna, in Chap. 46. Sect. 11 aforegoing, to which you must add of Sulphur dissolved **3** parts, of Arsenick one part (as afterwards is shewed) through all things doing, as in the place now cited is directed; and it will be a Medicine tinging every Body, and Mercury itself into true Sol, or better, according to the way now shewed. Read and peruse what we shall direct, and thereby you will be able to tinge to Infinity, if you have understanding, and erre not by the ambiguous sayings of the Philosophers.

XV. The Ferment of Sol for the Red. The Ferment of Sol is made of Gold, dissolved into its own Water (Aqua Regis) and decocted and prepared by the directions in Chap. 46. Sect. 16. aforegoing: So will you have the Ferment of Sol for the Red, which keep for use.

XVI. The Ferment of Ferments upon Mercury for the Red. *Dissolve Sal in its own water (which we shall hereafter teach) (i.e. Aqua Regis) to this Gold dissolved 1 ounce, add Sulphur 2 ounces, dissolved in the same Water together with it, Mercury 3 ounces, also disolved. Let all these be truly dissolved into most clear Water, which being mixt, decoct for one day, that they may be Fermented then draw off the Water 15 times, each time cohobating. Incerate with Yellow Virgins Wax, that is with half its Weight of Oyl of Blood, or Oyl of Eggs: then project up on crude Mercury, as you see requisite.* Here note, that if you perfect this Medicine, as we teach in our third Order, in Chap. 47, Sect. 21. 22, etc. following, of the Congelative Medicine of Mercury, you will find by Reiteration of the Work, and by Subtilization thereof, that one part, will tinge infinite parts of Mercury into most fine and high Gold, more Noble than any natural Gold whatsoever.

CHAPTER XLVIII

OF THE ALCHYMIE OF MERCURY.

I.   *Argent Vive,* which is also called *Mercury,* is a Viscous Water in the Bowels of the Earth, by most temperate heat United, in a total Union through its least parts, with the substance of White subtil Earth, until the humid be contemperated with the Dry, and the Dry with the humid equally. Therefore it easily runs upon a plain Superfices, by reason of its watery humidity, but it adhers not, although it has a Viscous humidity, by reason of the dryness, of that which Contemperatesit, and permits it not to adhere.

II.   This is also as some say, the matter of Metals with Sulphur, and easily adheres to three Minerals, *viz, Saturn, Jupiter* and *Sal,* but to Luna more difficultly, and to *Venus* more difficultly than to *Luna;* but to *Mars* in no wise but by Artifice. Hence you may collect a very great Secret. For it is amicable and pleasing to the Metals, and the Medium of conjoyning Tinctures; and nothing is submerged in Argent Vive, unless it is Sol. Yet Jupiter, and Saturn, Luna and Venus, are dissolved by it, and mixed; and without it, can none of the metals be gilded. It is fixed, and the Tincture of Redness of most exuberant perfection and fulgid splendor; and receeds not from the Commixtion,

till it is in its own nature. But it is not our
Medicine in its Nature, but it may sometimes help in
the Case.

III. Of the Sublimation of Argent Vive. This Work is
coin-pleated with its Terrestreity is highly purified,
and its Aquosity wholy removed. We remove it not by
adustion, because it has none, so the Art of
separating its superfluous Earth is to mix it with
things, where with it has not Affinity, and often to
reiterate the Sublimation from them. Of this kind is
Talck, and the Calx of Egg-shells, and Calx of white
Marble, as also Glass in most subtil Pouder, and every
kind of Salt prepared, for by these it is cleansed;
but by other things having affinity with it, (unless
they be bodies of perfection) it is rather Corrupted,
because all such things have a Sulphureity, which
ascending with it in Sublimation, corrupt it. And this
you may find to be true by Experience, because, when
you sublime it from Tin, or Lead, you find it, after
Sublimation, infected with blackness. Therefore its
Sublimation is better made by those things which agree
not with it; but it would be better by things with
which it does agree, if they had not Sulphureity.
Wherefore this Sublimation is better from Calx, than
from all other things, because that agrees little with
it, and has not Sulphureity.

IV. But the way of removing its superfluous aquosity, is, that when it is mixed with Calces, from which it is to be sublimed, it be well Gound and coxnmixt with them by Imbibition, until nothing of it appear, and afterwards the Wateriness of Imbibition removed by a most gentle heat of Fire, which receeding, the Aquosity of Argent Vive receeds with it; yet the Fire must be so very Gentle, as that by it, the whole substance of Argent Vive ascend not.

V. Therefore from the manifold reiteration of Imbibition, with Contrition, and gentle Aflation, its greater Aquosity is abolished, the residue of which is removed, by repeating the Sublimation often. And when you see it is most white, excelling Snow in its whiteness, and to adhere (as it were dead) to the sides of the Vessell; then again reiterate its Sublimation, without the feces, because part of it adheres fixed with the Feces, and can never by any Art or Ingenuity be separated from them. Or, afterwards, fix part of it as we shall teach you; and when you have fixed it, then reiterate Sublimation of the part remaining, that it may likewise be fixed.

VI. Being fixed, reserve it, but first prove it upon Fire; if it flow well, then you have administred sufficient Sublimation; but if not, add to it some small part of Argent Vive sublim'd and reiterate the

Sublimation till your end be accomplished: for if it has a Lucid and most white Color, and be porous, then you have well sublimed it; otherwise, not therefore in the preparation of it made by Sublimation, be not negligent, because such as its cleansing shall be, such will be its Perfection, in projecting of it upon any of the imperfect Bodies, and upon its own Body unprepared.

VII. Yet here note, that some have by it formed *Iron,* some *Lead,* others *Copper,* and others *Tin;* which happened to them through negligence in the Preparation; sometimes of *it alone,* sometimes of *Sulphur,* or of its Compeer *Arsenick,* mix with it. But if you shall by Subliming, directly cleanse and perfect this Subject, it will be a firm and perfect Tincture of Whiteness, the like of which is not in being besides.

VIII. Of the Coagulation of Mercury. Coagulation is the reducing a Liquid body to a solid Substance, by privation of the humidity: and is of Service. *I. For Indurating Argent Vive, which needs one kind of Coagulation. 2. For freeing dissolved Medicines from their watriness, which requires another. Argent Vive is* coagulated two ways: One by washing away its whole innate humidity from it: the other by Inspissation, till it be hardened, which is a laborious work. Some thought the Art of its Coagulation was to keep it long in a temperate Fire, who when they thought they had coagulated it, after removal of it from the Fire, found it to flow as before; whence they judged the work Impossible.

IX. Others, from natural principles, supposing that every humidity must necessarily by heat of Fire be converted into Dryness, indeavored by Constancy and perseverance, to continue the Conservation of it in the Fire, till some of them converted it, into a *White-Stone;* others into a *Red;* others into a *Citrine;* which neither had *Fusion* nor Ingress; for which cause they also cast it a way.

X. Others endeavoured to coagulate it with Medicines, but effected it not and so were deluded for that, 1. They either coagulated it not. 2. Or else it was insensibly extenuated. 3. Or the Coagulation was not in the form of a body: the reason of which things they knew not.

XI. Others compounding Artificial Medicines, coagulated it in projection; but that was not profitable, because they converted it into an imperfect Body, the cause of which they could not see. The reason and causes of these things therefore we think fit to declare, that the Artificer may come to the knowledge of his Art.

XII. Now, as the substance of *Argent Vive* is Uniform, so it is not possible in a short time, by keeping it constantly in a continued Fire to remove its *Aquosity;* so that too much haste was the cause of the first Error. And being of a subtile substance, it receeds from the Fire; therefore excessive Fire, is the cause of the Error of those Men, from whom it flies.

XIII. It is easily mixed with *Sulphur, Arsenick,* and *Marchasite,* by reason of Community in their Natures: therefore it appears to be Coagulated by them, not into the form of a *Body,* but of *Argent Vive* mixed with *Lead;* for these being fugitive, cannot retain it in the Contest of Fire, until it can attain to the nature of a *Body;* but through the Impression of the Fire, they fly with it; and this is the cause of the Error of them who so Coagulate.

XIV. Also *Argent Vive* has much humidity joyned to it, which cannot possibly be separated from it, but by Violence of Fire warily adhibited, with conservation of it in its own Fire; and they by augmenting this its own Fire as far as it can bear, take away the humidity of Argent Vive, leaving no part sufficient for *Metalick Fusion,* which being taken away it cannot be Melted, which is the cause of their Error, who coagulate it into a Stone not *fusible.*

XV. In like manner, *Argent Vive* has Sulphureous parts naturally mixt with it; yet some Argent Vive has more, some less, which to remove by Artifice is impossible. Now seeing it is the property of Sulphur mixt with *Argent Vive,* to create a *Red* or *Citrine* Color (according to its measure) the ablation of that being Made, the property of Argent Vive is by Fire to give a white Color. This is the cause of the variety of Colors, after its Coagulation into a Stone. Likewise it has the Earthiness of Sulphur mixt with it, by which all its Coagulations must necessarily by infected. And this the cause of the Error of those who coagulate it into an imperfect Body.

XVI. Therefore it happens from the diversity of the Medicines of its Coagulation, that divers bodies are Created in its Coagulation; and from the Diversity of that likewise, what is to be coagulated. For if either the Medicine, or that, has a Sulphur not fixed, the body created of it, must needs be soft: but if fixed the body must necessarily be hard. Also, if *White,* White; and if *Red,* Red; and if the Sulphur be remiss from *White* or *Red;* the *Body* likewise must be remiss; and if *Earthy,* the body must be imperfect; if not, not so. Also every not fixed Sulphur creates a Livid body; but the fixed, as much as in it lies, the Contrary: and the pure substance of it creates a pure body; the not pure, not so.

XVII. Also the same diversity doth in like manner happen in *Argent Vive* alone, without the Commixtion of *Sulphur,* by reason of the diversity of *Purifications* and *Preparations* of it in Medicines. Therefore an Illusion happens from the part of the Diversity of the Medicines; so that sometimes in the Coagulation of it, it is made *Lead,* sometimes *Tin,* sometimes *Copper,* sometimes *Iron;* which happens by reason of *Impurity.* And sometimes *Silver* or *Gold* is made thence, which must needs proceed from *Purity,* with consideration of the Colors.

XVIII. But *Argent Vive* is Coagulated by the frequent precipitation of it with Violence, by the forceable heat of strong Fire. For the Asperity of Fire easily removes its *Aquosity,* and this Work is best done by a Vessel of a great length, in the sides of which it may finde place to Coole and Adhere, and (by reason of the Length of the Vessel) to abide, and not fly till it can again be precipitated to the *Fiery Bottom* of the same; which must always stand very hot, with great Ignition: and the same precipitation be continued, till it be totally fixed.

XIX. It is also Coagulated, with long and constant retention in the Fire, in a Glass Vessell, with a very

112

long Neck, and round belly, the Orifice of the Neck being kept open, that the humidity may vanish thereby. Also it is coagulated by a Medicine convenient for it, which we will shew anon: which Medicine is of it, and is that, which most nearly adheres to it, in its profundity; and is commixed throughly in its least parts, before it can fly away. Therefore there is a necessity of collecting that, from things convenient to it, or agreeing with the same: Of this kind are all Bodies, also Sulphur, and Arsenick.

XX. But because we see not any of the *Bodies* in its nature to coagulate it; but that it flys from them, now neerly soever they agree together; we have therefore considered, that no Body adheres to it in its inmost parts. Wherefore, that Medicine must needs be of a more subtil substance, and more liquid fusion than Metals themselves are. Also by Spirits, remaining in their nature, we see not a Coagulation of it to be made, which is firm and stable; but fugitive, and of much infection. Which indeed happens by reason of the flight of the Spirits; but the other from the commixtion of the Adustible and Earthy substance of them.

XXI. Hence then it is manifestly evident, that from whatsoever thing the Medicine thereof is extracted that must necessarily be of a most subtil and most pure substance, of its own nature adhereing to it; and of liquefaction most easie, and thin as water; and also be fixed against the violence of fire. For this will coagulate it, and convert the same either into a *Solar* or *Lunar* nature: Studiously exercise yourself upon what we have spoken, and you will find the Mystery out.

XXII. But that you may not blame us, as if we had not sufficiently spoken thereof, we say, that this *Medicine* is extracted from *Metalick Bodies* themselves, with their *$ulphur,* or *Arsenick* prepared: Likewise from *Sulphur* alone, or *Arsenick* prepared; and it may be extracted from Bodies only. But from *Argent Vive* alone, it is more easily, and more nearly, and more perfectly found; because nature more amicable embraceth its proper nature, and in it more rejoyces than in any extraneous nature; and in it is a facility of extraction of the substance thereof, seeing it already hath a substance subtil in Act. Now the *ways* of acquiring this Medicine, are by *sublimation,* as is by us sufficiently declared: And the way of fixing it follows. But the way of Coagulating things dissolved, is by a Glass in Sand, with a temperate fire, until their aquosity vanish.

XXIII. *The way of fixing Argent Vive,* is the same with the way of fixing *Sulphur* and *Arsenick* and these waies differ not, unless that Sulphur and Arsenick cannot be fixed if their most thin inflamable parts, be not separated from them, with the subtil Artifice of dividing, by this ultimate way of fixation. But Argent Vive has not this consideration, therefore in this method, they need a greater heat than Argent Vive. In like manner they are diversified, because these (Sulphur and Arsenick) must be elevated higher by reason of their slowness than Argent Vive; and also because they require a longer time to be fixd in, and a longer Vessel for their fixation.

XXIV. Of the Medicine Coagulating of Argent Vive. It is taken from such matter, as the matter itself is (viz, as we have before declared) and that is, because Argent Vive, (seeing it is easily made to fly, without any Inflamation,) may suddenly adhere to it, in its profundity, and be conjoyned with it, in its least parts, and likewise inspissate, and conserve it in the fire by its own fixation, until it be better able to sustain the force of Fire, consuming its humidity; and convert it by the benefit of this, in a moment, into true Solisick and Lunisick, according to that for which the Medicine was prepared.

XXV. But seeing, we find not anything more to agree with it, then That, which is of its own nature, therefore by reason of this, we judged, that with

That, the Medicine thereof might be compleated; and we endeavoured by Art to make the Form of the Medicine agreeable to the same, viz. That it be prepared in the method and way now mentioned, with the instance of long continued labour; by which all the subtil and most pure substance of it, may be rendred perfectly White in Luna, but intensly Citrin in Sol.

XXVI.   Now this cannot be compleated, so as to create a Citrine Color, without the mixtion of a Thing tinging it, which is of its own nature. But with this most pure substance of Argent Vive, the Medicine is perfected by this our Art, which most nearly adheres to Argent Vive, and is most easily fluxed, and coagulates it, for it converts it into a true Solisick and Lunisick, with Preparation of that always preceeding.

XXVII. *The Grand Question is*, from what things this substance of Argent Vive may best be extracted? To which we Answer: It must be taken from those things in which it is: But according to Nature, it is as well in Bodies, as in Argent Vive itself, seeing they are found to be of one Nature: In Bodies more *difficultly;* in Argent Vive more nigh, or easily, but not more perfectly. Therefore of what kind soever the Medicine is to be, the Medicine of this Pretious Stone, must be as well sought in Bodies, as in the substance of

Argent Vive.

XXVIII. *But as to the Fixing of Argent Vive,* you must know, that it may be done, without being turned into Earth, and likewise fixed with conversion of it into Earth. For by hastening to its *fixation,* which is made by *precipitation* it is fixed and turned into Earth. Also by the successive *sublimation* of it often repeated, it is fixed likewise, and not changed into Earth, but gives Metallick fusion. This is manifest to, and proved by him who has experienced both fixations thereof, even to the Consummation of the Work; both by the hasty precipitation; and also by the flow, with continually repeated sublimations.

XXIX. This therefore is because it has a viscous and dense substance, the sign of which is the grinding of it by Imbibition, and mixtion with other things. For *Viscosity* is manifestly perceived in it, by the much adherency thereof. That it has a *dense substance,* he that has but one Eye, may manifestly see by its aspect, and by posing the vast Weight thereof. For while it is in its own Nature, it exceeds *Gold* in weight, being of a most strong Composition. Whence it is manifest, that it may be *fixed* without consumption of its humidity, and without conversion of it into Earth.

XXX. For by reason of the good adherency of parts, and the strength of its mixtion; if the parts of it be any wise inspissate by Fire, it permits itself no farther to be corrupted; nor suffers itself (by the Ingress of a furious flame into it) to be elevated into *fume;* because it admits not of Rarefaction, of its self, by reason of its density, and want of Adultion, which is made by combustible Sulphureity, which it hath not.

XXXI. Hence is seen; First, The Causes of the Corruption of *every of the Metals by fire,* which is, 1. From the Inclusion of a burning Sulphureity in the profundity of their substance, diminishing, them by Inflamation, and exterminating them also into fume, with extream consumption of whatever Argent Vive, is in them of good Fixation. 2. From a multiplication upon them, of an exterior flame, penetrating, and resolving them with itself into fume, of how great fixation soever, that which is in them is. 3. From the Rarefaction of them by Calcination, for that the flame or fire, does then penetrate into, and exterminate them. Therefore if all Causes of Corruption concur, such Bodies must needs be exceedingly corrupt: But if not all, the corruption is according to the number and proportion of the Causes which remain.

XXXII. Secondly, *The Causes of Goodness, and Purity of each Metal.* For seeing that Argent Vive, for no Causes of Extermination, permits itself to be divided into parts in its composition, (because it either with its whole substance receeds from the fire, or with its whole remains permanent in it) there is necessarily observed in it a cause of perfection: For it is that which overcomes Fire, and by Fire is not overcome, but it amicably rests, rejoycing therein, possessing Perfection, as we have found, with an Approximate Potency.

XXXIII. *Of the Putrifaction* of Argent Vive. It is cleansed two ways, either by sublimation, of which we have shewed the way already; or by way of a *Lavament,* of which the way is this. Put Argent Vive into a Stone, or Earthen Dish, and pour upon it as much Vinegar, as is sufficient to cover it: Set it over a gentle fire, and let it heat so far, as you may well hold your Fingers in it, and no more. Then stir it about with your Fingers until it be divided into most small Particles, in the similitude of Powder; and continue stirring it, until all the Vinegar be wholly consumed: After which wash-away the Earthiness remaining with Vinegar, and cast it away: Repeating this washing so often, till the Earthiness of the *Mercury* is changed into a most perfect Coelestine colour, which is a sign that it is throughly washed.

XXXIV. Of the Nature of Argent Vive. There is a

necessity of removing its Superfluities, for it has Causes of Corruption, viz. an Earthy substance, and an adustible watriness without Inflamation. Yet some have thought it to have no superfluous Earth and Uncleanness, but that is vain, and not true: For we see it to consist of much lividness, and not of whiteness; we see also a black and Feculent Earth, to be separated from it, with easie Artifice, by a *Lavation,* as abovesaid. But because we are by that to acquire a two—fold perfection, viz. 1. *To make a Medicine.* 2. *To perfect it.* Therefore we must necessarily prepare the same by the degrees of a two-fold *purification* for two cleansing of *Mercury,* are necessary. One by Sublimation for the Medicine, which shall be here shewed: The other by a *Lavament* for coagulation, which we have shewed at Sect. 33 above.

XXXV. For if we could make a *Medicine* of it, then there is a necessity to cleanse it from the foeculency of its Earthiness by *sublimation,* least it create a livid color in projection; and also to remove its fugitive watriness, lest it make the whole Medicine fugitive in projection, and to keep safe the middle substance thereof for *Medicine;* of which the Property is not to be burned, but to defend from combustion, and not to fly itself, but to make fixed, which is a perfection by manifold Experiences. For we see Argent Vive more nearly to adhere to *Arg-ent vive,* and to be more beloved by the same; but next to it *Gold* has place, and after that *Silver.*

XXXVI. Wherefore *hence* it follows, that *Argent Vive* is more friendly to its own nature; but we see other Bodies not to have so great conformity to, or unity with it; and therefore we find them in very deed, less to partake of the nature thereof. And whatsoever Bodies we see more to defend from adustion, those we judge to partake more of the nature of it; therefore it is manifest, that Argent Vive is the perfective and salvative from Adustion, which is the Ultimate of Perfection.

XXXVII. The second degree of its *Purification,* is for its *Coagulation:* And the washing away of its earthiness, for one day only is sufficient for it; the method of which washing we have largely declared, at Sect. 33. aforegoing: Being therefore so throughly washed, project upon it the Medicine of Coagulation, and it will be coagulated into a *Solifick* or *Lunifick* substance, according as the Medicine was prepared. From what is now said, it is manifest, that Argent Vive is not perfective in its nature; but that matter is, which is produced of it by our Art. And so likewise, it is in Sulphur and Arsenick. Therefore in these it is not possible to follow nature, but by our natural Artifice.

XXXVIII. It is also undeniably manifest that bodies containing the greatest quantity of *Argent Vive* are *Bodies* of perfection. Wherefore it is to be supposed, that those *bodies* are more nigh to perfection, which more amicable imbibe *Argent Vive.* The sign of this is the easie susception of *Ar gent Vive* by a *Solar* or *Lunar* body of Perfection. For this same reason, if a *body* altered do not easily receive *Argent Vive* into its Substance, it must needs be very remote from this perfection spoken of.

XXXIX. *The preparation of Argent Vive. Take of it one pound:* Vitriol Rubified, two pounds: Roch Alum Calcin'd, one pound: Common Salt, half a pound: Nitre, four ounces: Incorporate all together and sublime. Gather the white and Dense, and ponderous, which will be found about the side of the Vessel, and keep it for use. Now, if in the first Sublimation, you shall finde it Turbid or Unclean (which may be thro Carlesness) sublime it again, with the same Foeces, and reserve it as before.

XL. *The Regiment of Mercury.* It is done two ways. 1. You must Amalgamate it, well washed and purified as under directed.

2. You must Distill it and thence make an Aqua Vitae

or Spirit of Wine. The first way. Take *of Mercury 40 Ounces, of Sol, of Luna, of Venus, of Saturn, ana one Ounce, melt these bodies first the Venus and Luna, secondly the Sol, thirdly Saturn:*

Take all out of the Fire; having melted them in a large Crucible, and your Mercury in readiness, made hot in another: and when the said Metals begin to harden, pouer in the Mercury Leisurly, stirring the mixture with a stick, setting it again on the Fire, and taking it off, untill they be all amalgamated, with the whole Mercury. This Amalgama put to be dissolved for seven days, Extract the water with a Cloth, make the residue Volatile, giving Fire of Ignition. This again imbibe with its whole water, and put it to be generated, and again to be dryed for forty days, and you will finde a Stone, which put to be fixed, so will you have a Stone augmentable to Infinity. In this Book we have expounded all things which we have written in divers Books.

XLI. The Sublimation of Mercury. If you would perfectly sublime it, you must add to every pound of it, common Salt two pound and a half, Salt-Peter half a pound: mortify the Mercury wholly, grinding it all together with Vinegar, until nothing of the Mercury appear living in the mixture, then sublime it according to Art. It is a thing profitable.

XLII. *The Sublimation of Red Mercury. Take one pound of it, mix and perfectly grind it with Vitriol, Nitre, ana one pound, and sublime it from them Red and splendid.*

XLIII. Out of all that has been said it appears with evident Demonstration, that our Stone is procreated out of the substance of Argent Vive: But to unlock the Closure of Art, you must study to resolve Luna or Sol into their own dry water, which the vulgar call Mercury: And it is so, that a duodenary proportion (of the solutive water) may contain only one part of the perfect body. For if with gentle fire, you well govern these, you will find (in the space of 40 days) the body converted into mere water: and the sign of its perfect dissolution is blackness, appearing on its Superfices.

XLIV. But if you endeavour to perfect both Works, the White and the Red, dissolve each of the ferments by themselves, and keep them. This is Our Argent Vive extracted from Argent Vive, which we intend for Ferment. But the Paste to be fermented, we extract in the usual manner from imperfect bodies. And of this we give you a general Rule, which is, *That the White*

Paste is extracted from Jupiter and Saturn; but the

Red from Venus and Saturn: But every Body must be
dissolved by its self in the Ferment.

XLV. Sulphur we have proved is corruptive of every
kind of Perfection: But Argent Vive is perfective in
the Works of Nature, with compleat Regiment. So we,
not changing, but imitating Nature, (in Works
possible) do likewise assume Argent Vive in the
Magistery of this Work, for a Medicine of each kind of
Perfection, viz, both Lunar and Solar, as well of
Imperfect Bodies, as of Argent Vive Coagulable. And
seeing there is a two—fold difference of Medicines,
one of Bodies, but the other of Argent Vive truly
coagulable, we shall here discourse it.

XLVI. The matter per so, of this Medicine of every
kind is one only, already sufficiently known. Take
therefore that, and if you will work according to the
Lunar Order, learn to be expert in Operating, and
prepare that, with the known ways of this Magistery.
The intention of which is, That you should divide the
pure substance from it, and fixt part thereof, but
leave a part of cerating; and so proceeding through
the whole Magistery, till you compleat its desired
fusion. If it suddenly flows in hard Bodies, it is
perfect; but in soft Bodies, the contrary. For this
Medicine projected upon any of the Imperfect Bodies,
changes it into a perfect Lunar Body, if the known

Preparations have been first given to this Medicine: But if not, it leaves the same diminished, yet in one only difference of Perfection it perfects, as much as depends on the Administration of the Order of a Medicine of this kind. But this due Administration not preceeding, according to the third Order, it perfects in projection only.

XLVII. A Solar Medicine (of the Second Order) of every of the imperfect Bodies, is the same matter, and participates of the same Regiment of Preparation. Yet in this it differs, viz, in the greater subtilization of parts, by proper ways of digestion, and in the cominixtion of subtil Sulphur (under the Regiment of Preparation administred) with the addition of the matter now known.

XLVIII. The Regiment of it is the fixation of pure Sulphur, and the solution thereof: For with this the Medicine is tinged, and with it projected upon every of the Bodies diminished from perfection; it compleats the same in a Solar Complement, as much as depends upon a Medicine of the Second Order, the known and certain preparation of the imperfect body preceeding. Also the same projected upon Luna, perfects it much, in a peculiar Solar compleatment.

CHAPTER XLIX

THE INTRODUCTION TO THIS SECOND BOOK.

I. There are two things to be determined, viz, the Principles of this Magistery, and the perfection of the same. The Principles of this Art, are the Ways or Methods, of its Operations, to which the Artist applys himself in the Work of this Magistery:

These ways are divers in themselves: As, *1. Sublimation. 2. Descension. 3. Distillation. 4. Calcination. 5. Solution. 6. Coagulation. 7. Fixation, 8. Ceration.* All which we shall with much plainness declare.

II. The perfection consists 1. Of those things, and from the consideration of those things by which it is attained. 2. From the consideration of things helping. 3. From the consideration of the thing which lastly perfects. 4. And from that by which it is known, whether the Magistery was in perfection or not.

III. The consideration of those things by which we attain to the Compleatment of the Work, is the consideration of the Substance manifest, and of manifest Colors, and of the weight in every of the Bodies to be changed, and of those Bodies that are not changed, from the Radix of their Nature, without that Artifice: and the consideration of those likewise that are changed, in the Radix of their Nature by Artifice: with the consideration of the Principles of Bodies, according as they are profound, occult, or manifest; and according to their Natures, with or without Artifice.

IV. For if Bodies and their Principles, be not known in the profound or manifest properties of their Natures, both with and without Artifice, what is superfluous, and what is wanting or defective in them, cannot be known; and our not knowing those, would of necessity hinder us, from ever attaining to the perfection of their Transmutation.

V. The consideration of things helping Perfection, is the consideration of the Nature of those things, which we see adhere to Bodies without Artifice, and to make Mutation: And these are, *Marchasite, Magnesia, Tutia, Antimony, and Lapis Lazuli.* And the consideration of those which, without adherency, cleanse Bodies; such are *Salts, Allums, Nitre, Borax, Vitriol,* and other things of like nature, And the consideration of Glass of all sorts, and things cleansing by a like nature.

VI. But the consideration of the thing that perfects, is the consideration of chusing the pure Substance of

Argent Vive: and it is the Matter, which from the Substance of that, took beginning, and of which it was created. This Matter is not Argent Vive in its Nature, nor in its whole Substance, but it is part of it: nor is it now, but when the Stone is made for that illustrates and conserves from Adulation, which is a signification of Perfection.

VII. Lastly, The consideration of the thing, or certain Tryal and Examination, by which it is known, whether the Magistery be in Perfection or not; arises from the consideration of *1. The Cupel. 2. Cement. 3. Ignition. 4. Exposing it to the Vapours of Acid Things. 5. Extinction. 6. Commixtion of Sulphur burning Bodies. 7. Reduction after Calcination. 8. Susception of* Argent Vive.

All which with the former we declare, with their Causes from Experiences, by which you may certainly know, we have not erred.

CHAPTER L.

OF SUBLIMATION, VESSELS, FURNACES.

I. The cause of the Invention of Sublimation, was to unite Bodies with Spirits, (since nothing can possibly be united with a Body but a Spirit.) Or to find something that can contain in its self the nature both of Body and Spirit, which being cast upon bodies, (without being first purified,) either give not perfect Colors, or else totally corrupt, blacken, defile, and burn them, and this according to the diversitie of the same Spirit.

II. For Sulphur, Arsenick, and Marchasite, are burnings and wholly corrupt: Tutia (of every kind) burns not, yet gives an imperfect Color, 1. Because its adustive Sulphureity, which is easily inflamed and blackens is not removed. 2. Because its Earthiness is not separated: for Adustion may create a Livid Color, and Earthiness may form it.

III. These things therefore we are constrained to cleanse from their burning Sulphuriety or Unctuosity, and Earthy superfluity, and this can be done by no Artifice but by Sublimation: for when Fire elevevates, it makes ascend always the more subtile parts, leaving behind the more Gross.

IV. Hence it is manifest that Spirits are cleansed from their Earthiness by Sublimation, which Earthiness impeded Ingress, and gave an impure or diminished Color: from which being separated, they are freed from their Impurity, and are made more splendid, more pervious, and more easily to enter and penetrate the density of bodies, with a pure and perfect Tincture.

V. Adustion is also taken away by Sublimation; for Arsenick which before Sublimation was apt to adustion after Sublimation, will not be Inflamed, but receeds without Inflamation; the same you may find in Sulphur. And because in no other things than in Spirits, we saw an adherency to Bodies with Alteration, we were necessitated to make choice of them, and to purifie them by Sublimation.

VI. Sublimation then, is the Elevation of a Drything by Fire, with adherency to its Vessel but is done diversly according to the diversity of Spirits to be sublimed: for some are Sublimed with strong Ignition, others with moderate, and some again with a remiss heat of Fire.

VII. Arsenick, and Sulphur, are Sublimed with a remiss Fire; for otherwise, having their most subtil parts uniformly mixt and conjoyned with the Gross, their whole substance would ascend black or burnt, without any Purification: therefore you must find out the proportion of the Fire, and the Purification, with commixtion of the Feces or Gosser parts, that they may be kept deprest, and not suffered to ascend.

VIII. In Sublimation a threefold degree of Fire is to be observed. 1. One, so proportioned, as to make to ascend only the Altered, more pure, and Livid parts, till you manifestly see they are cleansed from their Earthly feculency. 2. Another degree is, that what is of the pure Essence remaining in the Feces, may be sublimed with greater force of Fire, viz, with IgnitIon of the bottom of the Vessel, and of the Feces therein, which you may see with your Eye. 3. The other degree is, a most weak Fire, which is to be given to the Sublimate without the Feces, so that scarcely anything of it may ascend, but that only which is the most subtil part thereof, and which in our work is of no value, for that it is a thing by help of which Adustion is made in Sulphurs.

IX. The whole intention therefore of Sublimation is, That 1. The Earthiness being removed by a due proportion of Fire. 2. And the most subtil and fumous part, which brings Adustion with Corruption, being cast away, we may have the pure Substance, consisting in Equality, of simple Fusion upon the Fire, and without any Adustion, or flying from the Fire, or Inflamation thereof.

X. Now that that which is most subtil is adustive, is evident, for that Fire converts to its own nature, all those things which are of affinity to it: it is of affinity to every adustible thing; and every thing the more subtil, the more adustible, therefore Fire is of most affinity to what is most subtle.

XI. The same is proved by Experience; for Sulphur or Arsenick not sublimed, are most easily inflamed, and of the two, Sulphur the more easily: but either being sublimed, are not directly inflamed, but fly away, and are extenuated without Inflamation, yet with a preceeding Fusion.

XII. Now the proof in the administration of Foeces, with their proportion, is, that such Matter be chosen, with which the Spirits to be sublimed may best agree, and wherewith they may be the more intimately mixed: for that Matter with which they are or may be most united, will be more potent in the retention of the Foeces of the Matter to be sublimed; the reason of which is evident.

XIII. But the addition of Foeces is necessary, because Sulphur or Arsenick to be sublimed, if they be not conjoyned with the Foeces of some fixed thing, would necessarily ascend with their whole substance not cleansed, which thing we know by experience to be truth: this is proved, because, if the Foeces be not perinixed with them thro' their least parts, then the same happens as if they had not Foeces, for their whole Essence will ascend without any cleansing.

XIV. Experience also proves this to be true, because when we sublime from a thing forraign to the nature of Bodies, we sublime in vain, so that they are found in no wise purified after the ascension: but subliming with the Calx of any Body, the sublimation is well, and with facility it is perfectly cleansed.

XV. The intention of Foeces then is, that they be administred or taken from the Calxes of Metals; for in them the work of sublimation is easie, but in other things most difficult; for which cause there is nothing than can be instituted in their stead; for that without the Calxes of Bodies, the Labor will be long, tedious, and most difficult, almost to desparation.

XVI. But in this there is some benefit, for what is sublimed without Foeces or the Calces of Bodies, is of greater quantity, but with Foeces of lesser: So also, what is calcined with the Calces of Bodies is of least quantity, but of easiest and most speedy Labor.

XVII. However every kind of Salt prepared, and things of like nature to it, excuses us from using the Foeces of Bodies, for that with them we make sublimation in a greater quantity; for separation of things to be sublimed from the Foeces, is easily made by solution of the Salts, which happens not in other things.

XVIII. But the proportion of Foeces is, that it be equal to the quantity of the matter to be sublimed, in which you cannot easily err: Yet if the Foeces be but half the weight, it may serve with care, to an experienced Man: For the less the Foeces are, the greater will be the Exuberation of the sublimate, provided, that according to the Subtraction of the Foeces, an abatement of the Fire be in proportion thereto: For in a small quantity, a small fire serves for perfection; in a great, a great; and in a greater quantity, a greater fire is required.

XIX. Now because fire is a thing which cannot be measured; therefore it is, that error is often committed in it, when the Artist is unskilful, as well in respect to the variety of Fornaces, as Woods and Vessels to be used, and their due joyning.

XX. Therefore in things to be sublimed, you must

remove their wateriness only, with a very finall Fire which being removed, if anything ascend by it, then in the beginning, this Fire must not be increased, that the most subtil part may (by this most weak fire) be separated, and put aside, which is the cause of Adustion.

XXI. But when little or nothing shall ascend (which you may prove by putting a little Cotton Weik into the hole in the top of the Aludel) increase the fire under it; and how strong the fire should be, the Cotton Weik will shew: For if little of the sublimate comes forth with it, or it be lcean, it shews your fire is small, and therefore must be encreased: But if much and unclean, that it is too great, and must be diminished.

XXII. When then you find your sublimate to come forth with the Weik Clean, and much, you have the due proportion of your Fire, but if unclean the contrary: For according to the quantity of cleanness, or uncleanness of the sublimate adhereing to the Cotton, must you order your Fire in the whole sublimation: by this means you may bring it to its due height without any error.

XXIII. Yet the way of Foeces is better, viz. To take Scales of Iron, or Copper calcined: these indeed by reason of the Privation of an Evil humiditity, do easily imbibe Sulphur or Arsenick, and Unite them with themselves; the method of which the experienced only know.

XXIV. It is fit therefore, that we should rightly inform you in the sublimation of these two Spirits (Sulphur and Arsenick) least you should erre through Ignorance: We say then, that if you put in many Foeces, and augment not the Fire proportionally, nothing of the Matter to be sublimed will ascend.

XXV. If you put in a small quantity of foeces, or none of the Calx of Bodies, and have not a fit proportion of Fire, the matter will ascend with its whole substance: So likewise by reason of the Fornace, you may err: For a greater Fornace gives a greater heat of Fire; a small Fornace, a small, if the Fewel and Ventholes be proportionate.

XXVI. If you sublime a great quantity of matter in a small Fornace, you cannot make a fire great enough for Elevation: If a small quantity in a great Forance, you will exterminate the sublimation by excess of heat. Again a thick Fornace gives a condensate and strong Fire: A thin Fornace, a rare and weak fire, in both which you may easily err.

XXVII. So also, a Fornace with large Vent-holes, gives

a clear and strong fire, but with small Vent-holes, a weak fire: And if the distance of space between the Fornace and the Vessel be large, the fire will be the greater, but if small, the less; in all which, without care, you may easily also err.

XXVIII. You must therefore build your Fornace according to the strength of the Fire you would have, viz, thick, with free Vent—holes, so as there may be a good distance between the Vessel, and sides of the Fornace, if you would have a great fire: But if a mean fire, in all these things you must find a mean proportion: All which we shall teach you.

GEBERS FORNACES.

Geber, lib. 2. cap. 50.

XXIX. If you would elevate a great quantity of matter to be sublimed, first be provided of a sublimatory of such a capacity, that it may contain your matter to be sublimed, the height of ones hand breadth above the bottom: To this fit your Fornace, so as the Aludel, or Sublimatory may be received into it, with the distance of two Fingers round about the Walls, or Sides of the Fornace; which being made, make also to it ten Vent-holes, in one proportion, equally distant, that there may be an equality of the fire in all parts thereof.

XXX. Then put a Bar of Iron into the Fornace transverse, which fasten at each end in the sides of the Forance, which Bar let be distant from the bottom of the Fornace about a Span, or 9 Inches: About an Inch above it the Sublimatory must be firmly placed, and inclosed round about to the Fornace.

XXXI. Now, if your Fornace can well and clearly dischange itself of the Fumosities, and the Flame can freely pass through the whole Fornace in the circuit of the Aludel, it is well proportioned; if not, it is not so. Then you must open its Vent-holes, and if by that it it is mended, all is well; if not, you must necessarily alter it, for the distance of the Vessel from the sides of the Fornace is too small: Wherefore enlarge the distance, and try it, continuing these Tryals, till it can freely quit itself of the smoak, and the flame is bright and clear.

XXXII. But as to the thickness of the Fornace, if you intend a great fire, it ought to be about 5 or 6 Inches; but if a moderate fire, 3 or 4 Inches; if a lesser fire, 2 or 3 Inches thick will be sufficient.

XXXIII. Then as to the Fewel, solid Wood gives a strong and durable fire: lighter Wood a weak fire, and soon ended; dry Wood gives a great fire and short; green Wood a small and long lasting. From the consideration of all these things, the diversity of Fires may easily be found out.

XXXIV. In the sublimation of Sulphur, the cover of the Sublimatory must be made with a great and large concavity within, after the manner of an Alembick with a Nose, for otherwise the whole sublimate may descend to the bottom of the Vessel, through too great heat, for that in the end of the sublimation, the Sulphur ascends not, unless with force of fire, even to Ignition of the Aludel: And if the Sulphur be not retained in the Concavity above, seeing it easily flows, it will descend again by the sides of the Vessel, to the very bottom, and nothing will be found sublimed.

XXXV. The Aludel is to be made of thick Glass, for other matter is not sufficient, unless it be thick, and of the like substance with Glass; because Glass only, or what is like to it, wanting Pores, is able to retain Spirits from flying away: For through Porous Vessels, the Spirits would pass and vanish.

XXXVI. Nor are Metals serviceable in this case, because Spirits (by reason of their Amity and Sympathy) penetrate them, and are united therewith: Therefore in the Composition of your Aludel, let a round Glass, or Concha, be made with a flat round bottom; and in the middle of the sides thereof, a Zone, or Girdle surrounding the same; and above the Girdle, cause a round Wall to be made, equidistant from the sides of the Concha, so that in this space, the sides of the Cover may freely fall without pressure.

XXXVII. But the height of this Wall (above the Girdle) must be according to the height of the Wall of the Concha, little more, or less. This done, let two Covers or Heads be made equal to the measure of this Concavity of the two Walls the length of the two Covers must be equal, and each a Span, or 9 Inches. The Figure of one of them also Pyramidal, in the superior parts of which Covers, must be two equal holes, one in each, so made that a Hens Feather may conveniently be put in.

XXXVIII. The intention of this Concha is, That its Cover may be moved at pleasure; and that the juncture might be ingenious, so that through it, though without any luting, the Spirits might not pass. But if you can better contrive this Vessel, you may do so, notwithstanding this our description.

XXXIX. Yet in this we have a special intention, that the interiour Concha, with its sides, should enter half way within its Cover, for seeing it is the property of Fumes to ascend, not to descend, by this means they are kept from vanishing: Also that the Head of the Aludel should be often emptied, lest part of what is sublimed (being over much) should fall down to the bottom again.

XL. Another intention is, that what ascends up in the form of pouder, near the hold of the head of the Aludel, be always kept apart, from that which is found to have ascended fused and dense in small lumps; porous and clear at bottom thereof, with adherency to the sides of the Vessel; for that it is known to have less of Adustion, than what is found to ascend nigh to the hold of the Head: Now the sublimation is well performed, if it be found clear and lucid, and not burnt with inflammation: This is the perfection of the subliming of Sulphur and Arsenick:

And if it be not so found, the Work must so often be repeated, till it is so.

CHAPTER LI

OF DESCENSION, AND THE WAY OF PURIFYING BY PASTILS.

I. There is a threefold Cause of its invention. 1. That when any matter is included in that Vessel, which is called, a Chymical Descensory, that after its fusion, it may descend through the Holes thereof, by which descent, we are assured, it has admitted a fluxing.

II. 2. That weak Bodies may be it be preserved from Combustion, after reduction from their Calces: For when we reduce weak Bodies from their Calces, we cannot reduce all their whole substance at one time: If then that part, which is first reduced into a body, should lie while the whole is reduced, a great quantity would vanish by the force of the Fire; so that it was necessarily devised, that one part so soon as it is reduced, may fall from the Fire, through this descensory.

III. 3. That the Depuration of Bodies might be so excellently performed, as to be freed from every extraneous thing: for the body descends in a Flux clean, and leaves everything which is alien thereto, in the Concavity thereof.

IV. Therefore as to the way or method thereof, we say, that the form of it must be such as its bottom may be pointed, and the sides of it without roughness, equally terminating in the aforesaid Acuity, or point of the bottom: And its cover (if any be needful) must be made in the likeness of a plain or flat Dish, and well fitted to it, and the Vessel with its Cover, must be made of good firm Earth, not easie to break or crack in the fire.

V. Then put in the matter which you would have to descend, upon round Rods or Bars made of like Earth, and so placed, as they may be more nigh the top than bottom of the Vessel. Then covering the Vessel, and luting the juncture, set it into the fire, and blow it until it is in Flux, and the whole matter descend into a subjacent Vessel.

VI. But, if the matter be difficult fusion, it may be put upon a Table plain, or of small Concavity, from which it may easily descend by inclining the head of the Descensory when it is in Flux; for by this means Bodies are purified.

VII. But they are yet better purified by Pastils, which method of Purification is of the same force, with the way of purifying by descension: For it holds the foeces of Bodies as well as a Descensory and better the way of which is thus.

VIII. Take the body which you intend to cleanse, and granulate it, or file it, or reduce it into a Calx, which is yet better, and more perfect: Mix it with some other Calx, which is not to be melted, and then make the body to flow.

IX. By this method, often repeated, Bodies are cleansed, but not with a perfect Mundification, which is to perfection; yet it a profitable purifying, that Bodies capable of perfection, may the better and more perfectly be transmuted.

X. For there is an Administration always to be before, and to proceed such a Transmutation, all which shall be declared in its proper place.

XI. The Descensory Fornace is made, as before described, and is wonderfully useful to the melting of Metals by Cineritiums and Cements. For all Calcined, Combust, Dissolved, and Coagulated Bodies, are reduced by this Fornace into a solid Mass, or Metal.

XII. *Cineritiums* also, and Cements, and Tests, or Crucibles, in which Silver is often melted, are put into this Fornace, for the recovering the Metals imbided.

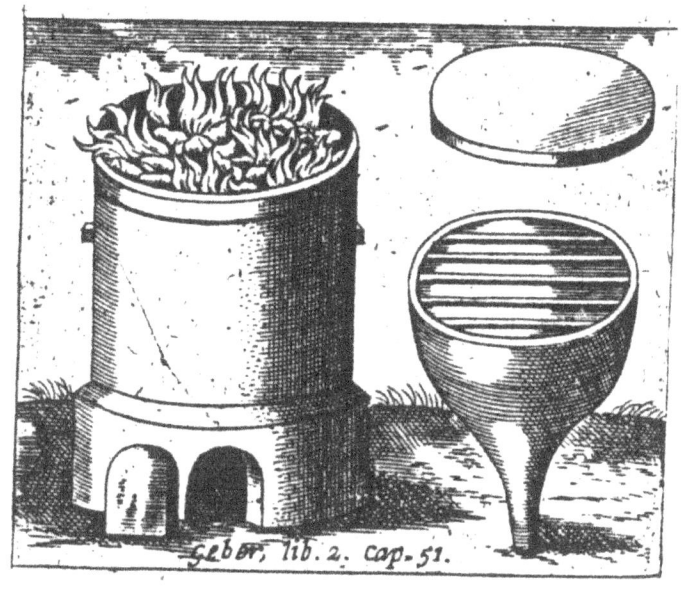

Geber, lib. 2. cap. 51.

CHAPTER LII

OF DISTILLATION, CAUSES, KINDS, AND FORNACES.

I. Distillation is the elevating of Aqueous Vapours in their proper Vessel; and is of divers kinds. 1. Either with fire, or without fire. Those made by fire is also two-fold. 1. Ascending by an Alembick. 2. Descending by a Descensory.

II. The Cause why Distillation was invented, was the purification of a liquid matter from its filth, and conservation of it from putrefaction. For we see things distilled (by what kinds soever of

Distillation) are made more pure, and more easily to be preserved from putrefaction.

III. But the special cause of Distillation by Ascent, or an Alembick, is the separating of a pure Water, without Earth or Foeces; for water so distilled has no feculency: And the Cause of the invention of such pure water, was for the Imbibition of Spirits, and of clean Medicines, left by the feculency of the Water, our Medicines, or Spirits might be defiled or currupted.

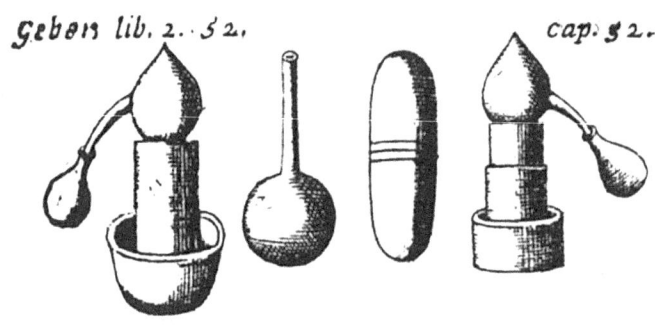

Gebers lib. 2..52.   *cap. 52.*

IV. But the cause of the Invention, which is made by Descent, or a Descensory, was the extracting its Oyl, pure in its Nature; because by Ascent, Oyls are not so easily had in their combustible Nature.

V. And the Distillation, which is made without fire, or by Filter, was invented for this cause sake, to clear water (whether distilled, or not distilled) from all manner of Impurities whatsoever.

VI. Distillation by Ascent is two—fold. 1. In Ashes, or Sand. 2. In Balneo, without Hay, or Wool in its proper Vessel, so disposed, that the Cucurbit, or Vesica may not be broken before the Work is finished.

VII. Distillation by Ashes or Sand, is done with a greater, stronger, and more acute fire: But that by Balneo, with a mild, soft, or gentle and equal fire; for Water admits not the Acuity of Ignition, as Ashes or Sand do.

VIII. Therefore by that Distillation which is made in Ashes, colours, and the more gross parts of the Earth are elevated; but by that in Balneo, the parts more subtil, and without color, and more approaching to the nature of simple Water, only arise. So that a more subtil separation is made by distillation in Balneo, than by a Distillation in Ashes or Sand.

IX. This is evident; for Oyl distilled by Ashes, is gross, thick, and foetid: But that being rectified in Balneo, the Oyl is separated into its Elemental parts; so that from a most Red Oyl, you have another most limpid, white, and serene, the whole redness remaining in the bottom of the Vessel.

X. By this Operation, we come to the determinate separation of all the Elements of every Vegetable; and of that which from Vegetables proceeds to a Being, and of every like thing. But by that which is made by Descent, we attain the Oyl of everything Vegetable, determinately, and of their like; and by Filteration we accomplish the clearness of every liquid thing.

XI. To Distil in Ashes. *You must have a strong earthen Pan, and fitted to the Fornace, like to the aforesaid Fornace of Sublimation, with the same distance from the sides of the Fornace, and with like Vent-holes; upon the bottom of which Pan sifted Ashes must be put to the thickness of one Fingers breadth (length almost) and upon the Ashes, the Retort, or Distillatory must be set, and covered round about with the same Ashes, almost as high as to the neck of the Alembick.* (Retort, or Distillatory.)

XII. *This done, put in the matter to be distilled, cover the Vessel with its Alembick, the neck of which must inclose the neck of the Cucurbit, or Vesica, lest what is to be distilled should fly away: Then lute the juncture, and begin the Distillation:*

But the Vesica, Cucurbit, Retort or Distillatory, with the Alembick Head, or Recipient, must be both of Glass, and the fire must be of strength, according to the exigency, or nature of the matter to be distilled, and to be continued till all that should be distilled is come off.

XIII. To distil in Balneo, is like the former, in a Cucurbit and Alembick; save that you must have an Iron or Brass Pot fitted to the Fornace: *Upon the bottom of the pot within, must be laid a Bed of Hay or Wooll, or other like matter, to the thickness of 3 Inches, that the Cucurbit may not be broken; and with the same the Cucurbit must be covered round about, almost as high as the neck of the Alembick, upon which lay sticks cross, and upon them stones, to hold the Cucurbit to the bottom of the Pot, and keep it firm and steady, that it be not raised by the Water, nor be broken by its moving up and down. Lastly, Put in Water till the Pot be full, which done, kindle the fire, and distil off the matter.*

XIV. To Distil by Descent. *You must have a Glass Descensory, with its Cover, and that put in which is to be Distilled, and then the Cover luted on, and fire made on the top, or over it that the Liquor may descend.*

XV. To Distil by Filtre. *Put the Liquor to be Distilled, into an Earthen, Stone, or Glass Concha, under which set another Vessel to receive the Distillation: The larger part of the Filter put into the Liquor, even to the bottom of the Concha, leting the narrower part hang over the side thereof, and over the under Vessel; so will the Liquor fall down through the Filter in the lower Vessel, without ceasing, to the last drop.* Where note, That if the Liquor be not clear enough the first time, it must be so often repeated, till it is as you desire it.

XVI. The Distillatory Fornace, is the same with the Sublimatory: But Fire must be administred according to the exigency of things to be Distilled: The way of doing which we have just now taught.

CHAPTER LIII

OF CALCINATION OF BODIES AND SPIRITS, WITH THEIR CAUSES & METHODS

I. Calcination is the bringing of a thing to Dust by Fire, through an abstraction of its humidity, holding the particles of the Body together.

II. The cause of the invention thereof, is, that the Adustive, corrupting and defiling sulphureity, may be abolished by Fire; and it is manifold, according to the diversity of the things to be calcined: for *Bodies* are calcined; and *Spirits* are calcined; as also other things foreign to these, but with a divers intention.

III. And seeing there are imperfect bodies of two kinds, *viz, Hard* as *Venus* and *Mars; and Soft,* as *Saturn* and *Jupiter;* all which are calcined; there was a necessity of calcining them with a several intention, viz., General and Special.

IV. They are calcined with one general Intention, when that their corrupting and defiling *Sulphureity* may be abolished by Fire: for every adustive *Sulphureity,* which could not be removed without Calcination, is thereby abolished from everything whatsoever.

V. And because the body itself is solid, and by reason of that solidity, the internal *Sulphureity* concealed within the continuity of the sustance of *Argent Vive,* is defended from Adustion; therefore it was necessary to separate the Continuity thereof, that the Fire coming freely to every its least parts, might burn the *Sulphureity* from it, and that the Continuity of *Argent Vive* might not defend it.

VI. The common intention also of CALCINATION, is the Depuration of the Earthiness; for it found that Bodies are cleansed by reiterated Calcination and Reduction as we shall hereafter show.

VII. Special Calcination is of *Soft Bodies,* and with these two intentions, that through it there may be an intention of Hardening and Fixing, which is accomplished by an Ignitious repition of Calcination upon them; and this is found true by Experience.

VIII. But why the Calcination of Spirits was invented, is, that they may be the better fixed, and the more easily dissolved into Water; for that every kind of thing Calcined is more fixed, than the not Calcined, and of easier solution: and because the Particles of the Calcinated, more subtilized by Fire, are more easily mixed with Water, and turned into Water.

IX. The Calcination of other things, is subservient to the Exigency of the Preparation of Spirits and Bodies, of which Preparation we shall speak more at large in the following: but these are not of Perfection.

X. The way of Calcination is divers, by reason of the diversity of things to be Calcined for *Bodies* are otherwise calcined than *Spirits* or *other things.* And *Bodies* divers from each other, are diversly Calcined. *Soft Bodies* have one general way, according to the intention, *viz,* That both may be Calcined by Fire only, and by the acuity of Salt prepared or unprepared.

XI. The first Calcination by *Fire* is thus: Have a vessel of *Iron* or *Earth,* formed like a Porringer, which let be very strong and firm, and fitted to the Fornace of Calcination, so that under it, the Coles may be cast in and blowed.

XII. Then cast in your *Lead* or *Tin* (the vessel being firmly set upon a Trivet of Iron or Stone, and fastened to the walls of the Fornace, with 3 or 4 Stones being thrust in, stiff between the Fornace sides and the Vessel, that it may not move: the form of the Fornace, must be the same with the Form of the Fornace of Great Ignition, of which we have spoken and shall speak more in the following).

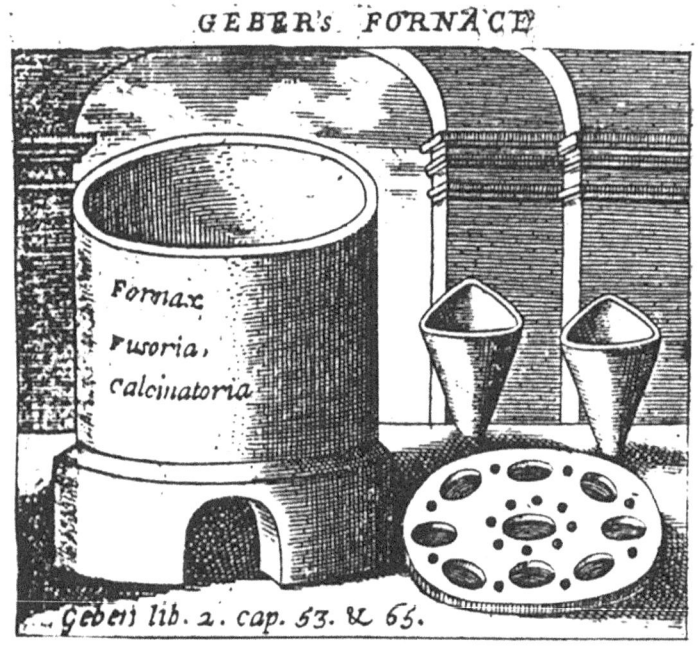

GEBER's FORNACE

Fornax
rusoria,
calcinatoria

Geberi lib. 2. cap. 53. & 65.

XIII. And the fire being kindled, sufficient for the fusion of the *Body* to be calcined, a skin will arise on the Top, which continually rake together, and take off with a Slice, or other fit Iron or Stone instrument, so long till the whole body is converted into Pouder.

XIV. If it be *Saturn,* there must be a greater fire till the *Calx* be changed into a compleat whiteness.

XV. Now understand that *Saturn* is easily reduced again into a Body from its Calx: but *Jupiter* with most difficulty: Therefore, be careful that you err not in exposing *Saturn* after its first Pulverization to too great a Fire and so reduce the *Calx* into a *Body,* before it is perfected: In this you must use temperance of Fire, and that leisurely augmented by degrees with great Caution, till it be confirmed in its Calx, and is not so easily reducible, but that a gentle fire must be given to the last compleating of the *Calx.*

XVI. Likewise be careful that you err not in *Jupiter,* by reason of its difficult reduction, for that intending to reduce it, you find it not reduced but a *Calx* still, or turned into Glass, and so then conclude its reduction impossible.

XVII. Now we say, that if a great Fire be not given in the reduction of *Jupiter,* it reduceth not: and if a great Fire be given, sometimes it reduces not, but possibly may be converted into Glass: the reason of which is, because Jupiter in the profundity of its nature has the fugitive substance of *Argent Vive* included: which if long kept in the Fire flies away; and leaves the Body deprived of humidity, so that it is found more apt to Vitrifie, than be reduced again into a metallic Body.

XVIII. For everything deprived of its proper Humidity, gives no other than a *Vitrifying Fusion,* whence it naturally follows, that you must hasten to reduce it with the speedy force of a Violent Fire; for otherwise it will not be reduced.

XIX. The Calcination of these Bodies by the Acuity of the Salt, is, the quantity after quantity of Salt be very often cast upon them in their fusion, and permixed by much agitation with an Iron Rod, while in fusion, till by the mixtion of the Salt, they be turned into Ashes: and afterwards by the same way of perfection, the Calces of them are perfected, with their considerations.

XX. But herein also is a difference in the Calces of these two Bodies: for *Lead* in the first work of Calcination is more easily converted into Pouder or Ashes than *Tin;* and yet the *Calx* is not more easily perfected than that of *Tin.* The cause of which diversity is, that *Saturn* has a more fixed humidity than *Jupiter.*

XXI. The Calcination of *Venus* and *Mars* is one, yet divers from the former, by reason of the difficulty of their Liquification. Make either of these Bodies into thin plates, heat them red hot, but not to Melting: for by reason of their great Earthiness, and large quantities of Adustive flying Sulphur, they are easily thus reduced into *Calx:* for the much Earthiness being mixed with the substance of *Argent Vive,* the due Continuity of the said *Argent Vive* is frustrated.

XXII. And thence comes their porosity, through which the flying Sulphur passes away, and the Fire by that means having access to it, Burns and Elevates the same; whence it comes to pass, that the parts are made more rare, and through discontinuity converted into Ashes.

XXIII. This is manifest, for that plates of Copper exposed to Ignition, yield a Sulphureous Flame, and make pulverizable Scales in their Superficies; which is done, because from the parts more nigh, a more easy combustion of the Sulphur must be made.

XXIV. The form of this Calcinatory Fornace is the same with the form of the Distillatory Fornace, save only, that this must have one great hole in the Crown of it to free itself from Fumosities: and the place of things to be Calcined must be in the midst of the Fornace, that the fire may have free access to them round about, but the Vessel must be of Earth, such as are Crucibles.

XXV. *The Calcination of Spirits.* You must give Fire to them gradually, and leisurely increase it, that they may not fly, till they be able to sustain the greatest Fire, and approach to Fixation: their Vessel must be round, every way closed, and the Fornace the same with the last mentioned. But you need not use greater Labour than what is to prevent their flight.

GEBER's FORNACES.

giber. lib. 2. cap. 53 55 56

XXVI. *Or thus,* As to the form of the Fornace, Let it be made square in length Four feet, and in breadth three Feet, *Luna, Venus* and *Mars,* or other things must be calcined in strong Dishes or Pans made of Clay, such as that of which Crucibles are made, that they may endure the strongest force of the Fire, to the total combustion of the matter to be Calcined.

XXVII. *Calcination is the Treasure of the Thing,* be not weary therefore, for imperfect Bodies are cleansed by it, and by reduction of the Calcinate into a solid Body or Mass of Metal again: then is our Medicine projected upon them, which is matter of Joy and Rejoycing.

XXVIII. *The Ablations of the Calces.* Have a large Earthen Vessel, full of pure, fresh hot Water, with this wash the *Calx* stirring it often, that all the Salt and Allom may be dissolved (with which they have been Calcined) then being settled, decant the Water gently: put the *Calx* again into hot Water and do as before, till it be perfectly washed, then dry and keep it for incineration.

XIX. *The Incineration of Calces washed. Take the former* Calx, *dissolve it in Spirit of Vinegar, 2 pounds of Common* Salt, Roch Allom, Sal Gem, ana 2 *ounces, in this water imbibe 4 ounces of the aforesaid dryed* Calx *till it has drank in all the said Water, then dry it and keep it for use.*

XXX. *The Reduction of Calces into a solid Mass. Take* the *former incinerated* Calx, *wash it with distilled Urine, till you have extracted all the* Salts *and* Alums, *with the filth of the Calcined Body, which being dryed, imbibe 4 pounds of this* Calx *with* Oyl *of* Tartar *1 pound, in 1 pound of which dissolve* Sal Armoniack 2 ounces, Salt Peter 1 ounce: *This Imbibition do at several times, drying and imbibing.Lastly dry it, and make it descend through a great descensory, and reduce it into a solid Mass, being purged from its Combustible Sulphureity by Calcination; and from its Terrestreity by its*

Reduction, so have you it purified from all accidental Impurities and defements, which happened to it in its Minera.

XXXI. But its mate foulness, which dwells in the Root of its Generation, must be obliterated or done away, with our Medicine, the greater part of which contains in itself the substance of *Argent Vive* according as the necessity of the Art requires.

XXXII. Again you must note, that Bodies are found to be of Perfection, if in the reiteration of their Calcination and Reduction, they loose nothing of their Goodness, in respect of Color, Weight, Quantity or Lustre, (of which great care is to be taken in the manifold reiterations of these OPerations) if therefore by repeating the Calcination and Reduction of altered Metals, they loose anything in their differences of Goodness, it is to be supposed you have not rightly persued the Art.

CHAPTER LIV
OF SOLUTION AND ITS CAUSE

I. Solution is the reduction of a dry thing into Water: and every perfection of Solution is compleated with subtile Waters such especially as are acute and sharp and Saline, having no Feces, as Spirits of Vinegar, of sower Grapes, of acid Pears, of Pomegranates, and the like distilled.

II. The case of this Invention, was the subtilization of those things which neither have Fusion nor Ingress, by which was lost the great advantage of fixed Spirits, and of those things which are of their Nature. For everything which is dissolved, must necessarily have the nature of *Salt* or *Alum,* or their like.

III. And the nature of them is that they give *Fusion* before their vitrification; therefore Spirits dissolved will likewise give Fusion: And since they in their own nature agree with Bodies, and each with other, Fusion being acquired, they must by that of neccessity penetrate *Bodies*, and penetrating them, transmute them.

IV. But they neither penetrate nor transmute without

our *Magistery* or Art, viz, That after Solution and Coagulation of the Body, there be added to it some one of the Spirits purified, not fixed; and then to be so often sublimed from it, till it remains with it, and gives to it a more swift Fusion, and conserves the same in *Fusion* from *Vitrification.*

V. For the nature of Spirits is not to be vitrified, but to preserve the mixture from Vitrification, as long as they are in it: Therefore the *Spirit* which more retains the nature of *Spirits,* defends or preserves from Vitrification: And a *Spirit* only purified, more preserves than a *Spirit,* purified, calcined, and dissolved: Therefore there is a necessity of mingling such a Spirit with the body; for from these there results good *Fusion* and *Ingress,* and true *Fixation.*

VI. Now we can demonstrate by natural operation, that things only holding the Nature of Salts, Alums, and the like, are soluble: For in all nature we find no other things to dissolved but them; therefore, what things soever are dissolved, must of necessity be dissolved by their nature or property.

VII. Yet since we see all things truly calcined, to be dissolved, by reiteration of Calcination and Solution; therefore we by that prove, that all Calcinates approach to the nature of Salts and Alums, and must of necessity be themselves, attended with these properties.

VIII. The way of Solution is two fold: By hot Dung, and by boiling or hot water; that is, in *Balneo;* of both which there is one intention and one effect.

IX. To dissolve by Dung, is, That the Calcinate be put in a Glass Vessel, upon which must be af fused Spirit of Vinegar, or the like, double its weight: Then the mouth of the Vessel must be so closed, or stopt, that nothing may go forth, and the matter with its Vessel set in hot Dung to be dissolved, and the Solution afterwards filterated.

X. But that which is not yet dissolved, must be again calcined, and after Calcination, in like manner dissolved, until by repeating the labour, the whole be dissolved as before which also filter.

XI. The way of dissolving by boiling water is more speedy, thus: Put the Calcinate in like manner into its Vessel, with Vinegar poured on it as before; and the mouth being well closed, that nothing expire, set the Vessel, buried in Straw, into a pot full of water, as in Distillation in *Balneo,* then kindling the fire, make the water boil for an hour: which done, decant the Solution and filtrate.

XII. And that which is undissolved, let it again be calcined; and then again in the same manner dissolved; which Work so often repeat, till the whole is finished.

XIII. The Dissolutory, or dissolving Fornace, is made with a pot full of water, with Iron Instruments, in which other Vessels are artificially retained that they fall not: These are the Vessels in which every Dissolution is made.

geber, lib. 2. cap. 54.

XIV. Bodies are in a two fold way brought to perfection, either (1) By the way of Preparation or (2) By commixtion of *Perfect Bodies* with the *Inperfect, i.e.,* by Medicines prepared for the purpose.

XV. Now we say that the *Body* cleansed by the way of Calcination (as afore said) and Reduced, must either be filed or Granulated thus; being melted, we pour it upon a Table-Board full of small holes, over cold water, the water being well stirred while this is doing.

XVI. The body thus granulated, we put into our *Dissolving Water* (*or* AF., made of *Nitre* and *Vitriol)* as to one halt thereof; or dissolve the filings of the same body in the said AF. into a limpid water; then add to it of Ferment prepared, to a third part of its own weight: Abstract the water and revert, or, cohobate it, and repeat this 7 times. After it is reduced into a *Body*, prove it in its *Examen,* and you will rejoyce for the Treasure you have found.

XVII. And because we have treated of the perfect *administration* of *Imperfect Bodies,* we should now give you the special,true and certain Rule for every particular body; but that being done already for *Saturn, Jupiter, Mars, Venus* and *Luna,* in their respective Chapters aforegoing, where we treat of their Regiment, we shall refer you thither.

XVIII. Mercury *also purified and fixed,* has power to take off or away the foulness of imperfect Bodies, and

167

to brighten them or illustrate them. And *Fixed Sulphur* extracted from bodies, to tinge or color them with splendor. Hence you may learn a great Secret, viz, *That* Mercury *and* Sulphur *may be extracted as well from imperfect bodies rightly prepared, as from the perfect.* Purified Spirits also, and middle Minerals, are a great help, and very peculiar, for bringing on the Work to perfection.

XIX. The Dissolving Water or AF. *Take* Cyprus Vitriol 1 *Pound,* Sal— Nitre, half *a pound,* Roch—Alum, *a fourth part, Distill off the water with a red hot heat, for it is very solutive; and use it as we have before in several places taught.* This may be made more acute.

CHAPTER LV
OF COAGULATION AND ITS CAUSES.

I. *Coagulation* is the Reduction of a thing Liquid to a solid substance, by deprivation of its moisture; for which there is a two—fold Cause; one is the Induration of Hardening of *Argent Vive* (of which we have already treated, Chap. 48, Sect. 8, ad 23).

II. The way of Coagulating things dissolved, is by a Glass placed in Ashes up to its Neck, and an equal Fire, not too hot put under it, and to be continued till the whole Aquosity is Vanished.

III. Now seeing it is not possible to remove the true Essence of anything in nature, the thing itself remaining, therefore it is said to be impossible to separate these corrupt things from them: for this cause some Philosophers have thought this Art not poss-ible to be attained, and *We,* and indeed other *Searchers* in this Science have been brought to this very State of belief.

IV. By reason of this, we as well as they, were driven to amazement, and for a long space of time lay under the shade of Desparation, yet returning to ourselves, and being perplexed with the immense trouble of dispairing thoughts and meditations, we considered

Bodies diminished from Perfection, to be foul in the profundity of their Nature, and nothing pure or clean to be found in them, because it was not in them according to Nature; for that which is not in a thing cannot be found there.

V. Seeing then nothing of perfection is found in them, therefore necessarily also, in the same nothing superfluous remains to be found, in separation of the divers substances in them, and in the profundity of their Nature, therefore by this, we found somewhat to be diminished in them, which must necessarily be compleated, by matter fit for it, and repairing the defect.

VI. Dimunition in them is the Paucity of *Argent Vive,* and not right Spissation or Coagulation of the same, therefore to compleat them, you must sufficiently augment the *Argent Vive:* Then rightly Inspissate or Coagulate; and lastly induce a permanent fixation (of which we shall speak in the next Chapter).

VII. But this is performed by a Medicine created of that: And this Medicine when brought forth into being from *Argent Vive,* by the benefit of its brightness and splendor, it hides and covers their Cloudiness, draws forth their Lucidity, and converts the same into Splendor, Brightness and Glory.

VIII. For which *Argent Vive* is prepared into a Medicine, and cleansed by our Artifice; it is reduced to a most pure and bright Substance, which being projected upon Bodies wanting of perfection, will illustrate or Tinge them, and by its fixing power perfect them: which Medicine we declare in its due time and place.

CHAPTER LVI
OF FIXATION AND ITS CAUSES.

I. Fixation is right disposing a Volatile or Fugitive thing to abide and endure in the fire: The cause of the Invention thereof is, that every Tincture, and every Alteration may be perpetuated in the thing altered and not vanish.

II. It is manifold according to the diversity of things to be fixed, which are all the Bodies diminished from perfection, as *Saturn, Jupiter, Mars* and *Venus;* and according to the diversity of Spirits also, which are *Sulphur and Arsenick in* one degree, and *Argent Vive* in another; Also *Marchasite, Tutia, Magnesia,* and such like in the Third.

III. Therefore, those bodies diminished from perfection, are fixed by their Calcination, because thereby they are freed from their volatile and corrupting Sulphureity; the which we have sufficiently declared in the Chapter of Calcination. Also the manifold repititions of sublimation, more swiftly and better do abbreviate the time of the Fixation.

IV. For this cause there was a second way of fixation found out, which is by precipitating of it, sublimed into heat, that it may constantly abide therein until it be fixed.

V. And this is done by a long Glass Vessel, the bottom of which (made of Earth, not of Glass, for that it would crack), must be artificially connexed with good luting; and the ascending matter when it adheres to the sides of the Vessel, must with a spatula of Iron or Stone be thrust down to the heat at bottom, and this precipitation repeated until the whole matter be fixed. How *Sulphur, Arsenick, Argent Vive, Marchasite, Magnesia and Tutia,are* to be fixed, we have taught in their proper Chapters aforegoing.

VI. *The Fixatory Fornace or Athanor.* It must be made after the manner of the Fornace of Calcination, and in it must be set a deep Pan full of ashes. But the Vessel, with the matter to be fixed, being firmly sealed, must be placed in the middle of the Ashes, so that the thickness of the Ashes underneath, and above in the compass of the Vessel, may be about four inches, or according to that which you desire to fix: Because in fixing *One,* a greater fire is required than in fixing *another.*

VII. By this Fornace, and this way the Ancient Philosophers attained to the work of the Magistery; which to men truly Philosophizing may be easily known, from what we have more than enough demonstrated in these our Books. And by those especially who are real searchers out of the Truth; we have given you the figure of the *Athanor,* yet let not this stop your further invention, if you can possibly find out anything more fit and ingenious nearly; but more nearly in *Argent Vive:* Whose humidity we see not to leave their Earth, by reason of the strong union which they have, and which nature has bestowed upon them in the Work of their Mixture.

VI. But in all other things having humidity, you may find by experience, that the same is separated in Resolution from their Earthy substance; and after separation thereof, that they are deprived of all humidity: In Spirits aforesaid, it is not so; so that we cannot omit taking them into the Work of *Ceration.*

VII. The way of *Ceration* by them is thus: *You must sublime them so often upon the thing to be* Cerated, *until remaining with their humidity in it, they give good Fusion:* But this cannot be effected before the perfect cleansing of them from every Corrupting thing.

VIII. And it seems better to me that these should be first fixed by *Oyl of Tartar,* and every *Ceration,* fit and necessary in this Art be made with them.

IX. Our PHILOSOPHIC CERATIVE WATER is thus made: Take *Oyl distilled from the White of Eggs: Grind it with half so much of* Sal Nitre, *and* Sal Armoniack, ana, *and it will be very good. Or, Mix it with* Sal Alkoli, *and distill as before.* And the more you reiterate this labour, the better it *incinerates. Or, Conjoyn the aforesaid Oyl, with Oyl of Tartar, and thence Distil a White* Incinerative Oyl.

X. *A Red Incinerative Oyl* is thus made: Take *Oyl of Yolks of Eggs, or of Humane Hair, to which adjoyn as much* Sal Armoniack; *mix and distil: Repeat this distillation three times, and you will have a most Red Incinerative Oyl.*

CHAPTER LVII

OF CERATION AND ITS CAUSE.

I. *Ceration* is the mollification or softening of and hard thing not fusible, unto Liquification: Whence it is evident, that the cause of the Invention of it was, That the matter which had not ingress into the Body for Alteration (by reason of Privation of its Liquefaction) might be softened, so as to flow, and have ingress.

II. Wherefore some thought Ingress was to be made with liquid Oyls and Waters, but that is error, and wholly remote from the Principals of this Natural Magistery, and denied by the manifest Operations of Nature.

III. For we find not, in those Metallick Bodies, that Nature has placed and humidity soon, or easy to be taken away, but rather one of long duration, for the necessity of their Fusion and Mollification: For had they been replenished with an humidity easie, or soon to be removed, it would necessarily follow, that the *Bodies* would be totally deprived of it, in one only Ignition, so that none of the Bodies could afterwards be either hammered or melted.

IV. Therefore, imitating the operations of Nature, we follow her way in *Cerating*. Nature Cerates in the

Radix of fusible things, with an humidity, which is above all humidities, and able to endure the heat of fire: Therefore it is necessary for us also to *Cerate* with the like humidity.

V. But this Cerative Humidity is in nothing better, more possibly, or more nearly found than in these, *viz,* in *Sulphur,* and in *Arsenick.*

XI. Oyl of Verdigrise is thus made: Dissolve Verdigrise in water of Sal Armoniack, with the same Coagulated, mix Qyl of Eggs, anddistil the mixture, which Distillation repeat thrice; so shall you have Oyl of Verdigrise, fit and profitable for Incineration.

XII. *Oyl of Gall;* its made by distilling an oyl from the Gall, as from human Hair; doing in all things as in the former.

XIII. I do not say that these Oyls can give a Radical Mineral Humidity, as in *Sulphur* and *Arsenick,* but they preserve the Tincture from Combustion, until it enters, or makes an Ingress; and afterwards they fly in the Augmentation of the fire.

XIV. After the matter is *Incinerated,* it may be necessary to melt it, which you must do in a *Fusory,* or *Melting Fornace.* This *Fornace* is that in which all Bodies are easily melted by themselves: It is a *Fornace* much in the use among *Melters of Metals:* Also *Aurichalcum* is melted in this Fornace, and tinged with *Tutia* or *Calaminaris* as is known to such as have made a Tryal.

CHAPTER LVIII

THAT OUR MEDICINE IS TWO-FOLD, ONE FOR THE WHITE & ONE FOR THE RED. YET THAT WE HAVE ONE ONLY MEDICINE FOR BOTH, WHICH IS MOST PERFECT.

I. We Demonstrate that Spirits are more assinuated to Bodies, than any other thing in nature; for that they are more United, and more friendly to Bodies, than all other things; so that we affirm, that these alterations of Bodies in the first Invention, are their true Medicines.

II. And as we have been exercised in all kinds, in the transformation of imperfect Bodies, with firmutation into a perfect Lunar and Solar Body, so we find that the Medicine for them must be divers according to the intention of the Bodies to be transmuted.

III. And since Metals to be transmuted are of a two-fold kind, viz. Argent Vive Coagulable in Perfection, and Bodies diminished from Perfection: and these again manifold, some being hard, sustaining Ignition, as Mars and Venus; others soft, not enduring it as Saturn and Jupiter; the Medicine perfective must also be necessarily manifold.

IV. And altho Mars and Venus be of one kind, yet they

differ in a certain special property, the one being not Fusible, the other fusible; therefore Mars is perfected with one Medicine, and Venus with another: The first indeed is totally unclean, but the other not: the former has a Dull whiteness; the latter that of Redness and Greenness: all which force a necessity of a Diversity in the Medicine.

V. Also the soft Bodies, Saturn and Jupiter, seeing they less differ, do necessarily require also a Divers Medicine: the first of them is indeed Unclean, the latter Clean; and they are all rendered more Mutable, now made Lunar than Solar Bodies; therefore the Medicine for each of them must be two-fold: One White, changeing into a White Lunar Body: and one Citrine, changeing into a Citrine Solar Body.

VI. Since then in every of the Imperfect Bodies is found a two—fold Matter, Solar and Lunar; the Medicines perfecting all Bodies, will be in number Eight.

VII. So also Argent Vive is perfected into a Lunar and Solar Body; therefore of the Medicine altering or perfecting it, there is a two-fold difference: so that all the Medicines which we have invented, for the Compleat alteration of every imperfect Body, will be in number Ten.

VIII. However, with constant and continued Labor, and great search and invention, we have been desirous to exclude the Use of these Ten Medicenes, by the Invention and advantage of One Only Medicine: and with our long and very Laborious search, by certain Experience, we have found One Medicine, by which the hard was softned; the soft Body hardned; the fugitive fixed, and the Soul illustrated with Splendor or Brightness ineffable, and beyond Nature.

IX. Notwithstanding, it is here expedient, that we should particularly speak of all these Medicines with their Causes, and the evident experiences of their probations. We will first then declare the series of the Ten Medicines, fitted to all the Bodies, then to Argent Vive, and lastly proceed to the Medicine of the Magistery, perfecting all Bodies; yet with the preparation imperfect Bodies need.

X. And least we should be carped at by the Envious, as Writing an insufficient Treatise of Art, We here first of all present the preparation of all the imperfect Bodies, assigning the Causes of the necessity thereof, by which (in Our artifice) they are made apt to receive the Medicine of Perfection, in every degree of Whiteness and Redness, and to be perfected by the same: and after these a Narration of all the Medicines before mentioned, themselves. The Preparations of Saturn, Jupiter, Mars, Venus, and Argent Vive here mentioned. See Chap. 42. Sect. 14. ad 20. Chap. 43. Sect. 11. Chap. 44. Sect. 12, 13, 14. Chap. 45. Sect. 12, 13. Chap. 48. Sect. 33. The preparation of the Medicines, see Chap. 44. Sect. 15, 16, 17. Chap. 45. Sect. 18 ad 23. Chap. 46. Sect. 6. Chap. 48. Sect. 33. etc.

XI. From what has been said, 'tis evident, that what Nature left Superfluous or deficient in every of those Bodies that are imperfect; has been in part declared: and since it happens that the mutable Bodies of Imperfection, are of a two-fold kind, viz. soft and Ignible, as Saturn and Jupiter; and hard and not fusible with Ignition, as Mars and Venus, the first indeed not fusible, but the other fusible with Ignition; Nature has taught us, That according to the diversity of Essences in the Radix of their Nature, divers Preparations, according to their Wants, must be administred to them.

XII. There are two Bodies of Imperfection of one kind, viz. Lead, which is Black, or Saturn; and Tin, which is White, or Jupiter; which from the innate Root of their nature, are divers each from other, in the profundity of their hidden parts, as well as in those which are outward.

XIII. For Saturn is cloudy, livid, ponderous, black, without stridor or crashing, totally mute: But Jupiter is white, a little livid, crashing much, a little sounding, and something bright; Of the Differences of which we have already spoken in their particular Chapters aforegoing.

XIV. From which Causes of Difference, according to more and less, you must collect the order of the Preparations; wherein we have shewed, first, The Preparation of Bodies; afterwards of Argent Vive coagulable. Now in the preparation of Bodies, nothing of Superfluity is to be removed from their profound, or inward Parts, but rather from their manifest or outward.

CHAPTER LIX

OF THE MEDICINE, TINCTURE, ELIXIR OR STONE OF THE PHILOSOPHERS, IN GENERAL.

1. The five different Properties constituting this Medicine.

I. Unless everything superfluous be taken away, either by Medicine or preparation from imperfect Bodies, viz. Every superfluous Sulphureity, and every unclean Earthiness, they cannot be purified, so, as that in Fusion they be not separated from the Commistion after projection of the Medicine altering them: when you have formed this you have found one of the five differences of perfection.

II. Also, if the Medicine do not illustrate, and alter and alter into a White or Citrine Color (according to what your intention is) inducing a splendent brightness, and admirable Lucidity; Bodies diminished from perfection are not perfected to the utmost.

III. So also, if it abides not Lunar or Solar Fusion, it is not changed into perfection; because it abides not in the Tryal; but is altogether separated, and receeds from the Commixtion; which you may more amply determine by the Cineritium, of which we shall speak hereafter.

IV. If likewise the Medicine be not perpetuated with a firm alteration, so that the Impression of Tincture, and Finity is not permanent but vanished in the Fire upon probation.

V. If it attains not to the weight of Perfection, (having the true ponderosity of Luna and Sol,) it is not firmly changed to a perfect compleatment of Nature: for this weight is one of the signs of perfection. Seeing therefore these differences of perfection are five, there is a necessity that our Medicine should exhibit these Differences in Projection. Also it is evident from hence, That this Medicine must be prepared from Things having Affinity to Bodies, readily altering, and amicably adhering to them in their profundity: But searching through Universal Nature, we have found nothing which can do all this so well as Argent Vive prepared, according to our Directions, of which the true Medicine is made to the highest Perfection.

2. The Preparations of the Medicine, that it may give the aforesaid different Properties.

VI. Now since it changes not, without the alteration of its Nature, therefore it ought necessarily to be prepared, that it may be mixed even in the profundity of Bodies, viz. That its substance may be made such, that it may be mixed even in the profundity of the Body alterable, without separation forever.

VII. But this cannot be done, without it be very much subtilized with certain and determinate sublimations, as we have taught in Chap. 48. Sect. 3, 4, 5, 6, 7. aforegoing: Likewise its Impression cannot be permanent, unless it be fixed, nor can it illustrate, unless its most splendid substance be extracted from it according to Art, with a fit fire.

VIII. Nor can this Medicine have perfect Fusion, unless great Caution be used in its fixation, that it may soften hard Bodies, and harden the soft. And it can only do that, when a sufficiency of its humidity is preserved, proportionate to the necessity of the Fusion desired.

IX. Whence it is evident, that it should have such a Preparation as may make it a most sulgent and purely clean substance, and fixed also; but these things must be done with such great Caution, (in respect to the regulation of the fire, and way of fixing) that in removing its Humidity, so much may be still left, for compleat and perfect Fusion.

X. If by this Medicine, you would soften Bodies hard of Fusion; in the beginning of its Preparation, a gentle fire must be adhibited: For a soft fire is Conservative of Humidity, and Perfective of Fusion.

XI. There is also many other Considerations of the Weight, with their Causes and Order. This Cause of great weight, is, the subtilty of the substance of Bodies, and uniformity in their Essence: By which the parts of them may be so condensed, that nothing can come between. And the Density of Parts, is the encrease of weight, and the Perfection thereof.

3. The Six Properties of things from which the Medicine is extracted.

XII. First, They have in themselves an Earth most subtil and incombustible, altogether fixed with its own Proper Radical Humidity, and apt for fixing.

XIII. Secondly, They have an airy and fiery Humidity, so uniformly conjoyned to that Earth, that if one be Volatile, so is the residue: And this same Humidity abides the fire beyond all Humidities, even to the cornpleat termination of its own Inspissation, without Evaporation, inseparable from the Earth annexed to it, with a compleat permanency.

XIV. Thirdly, The Disposition of their Natural Humidity is such, that by help of its own Oleaginity in all differences of its Properties, it contemperates the Earth annexed to it, with such an Unctuosity, and with such a Homogene and equal Union, and bond of inseparable Conjunction, that after the degree of final Preparation, it gives a good Fusion.

XV. Fourthly, The Oleanginous Property, is of so great purity of Essence, and so artificially cleansed from all Combustible matter, that it burns not any Bodies with which it is conjoyned through their least parts, but preserves them from Combustion. Hermes. Chap. 12. Sect. 5. aforegoing.

XVI. Fifthly, It has a Tincture in itself so clear and splendid, White, or Red, clean and incombustible, stable and fixed, that the fire cannot prevail against it to change it: Nor can Sulphurous, Adustive, or Sharp, Corroding Bodies, Corrupt and Defile the same.

XVII. Sixthly, The whole Compositum, incerated with its final Compleatment, is of so great Subtilty and Tenuity of Matter, that after the end of its Decoction, it remains in Projection of most fusion like water, and is of profound Penetration, to the greatest perfection of the Body to be Transmuted, how Fixed so ever it be; adhering thereto with an inseparable Unity or Conjunction, against the force of the strongest Fire; and in that very hour, by virtue of its own Spirituality, reducing Bodies to Volatility.

4. *The Seven Properties of the Medicine itself.*

XVIII. First, Oleaginity, Giving in Projection Universal Fustion, and Diffusion of the Matter; For the first thing after Projection of the Tincture, is the sudden and due Diffusion of the Medicine itself, which is perfected and rendered Viscous, with a Mineral Oleaginity.

XIX. Secondly, Tenuity of Matter, or the Spiritual substance thereof, flowing very thin in its Fusion, like Water, Penetrating to the Profundity of the Body to be Transmuted, for that immediately after Fusion, the Ingression thereof is necessary.

XX. Thirdly, Affinity, or Vicinity, between the Elixir or Tincture, and the Body to be Transmuted, giving adherency in Obviation and Retention of its like; because immediately after Ingress of the Medicine, Adherency is convenient and necessary.

XXI. Fourthly, Radical Humidity, Fiery, Congealing, and Consolidating the Parts retained, with adherence, to what is Homogene to it, and the union of all its said Homogene parts, inseparably forever. Because after Adherency, Consolidation of the parts by a Radical Viscous Humidity is necessary.

XXII. Fifthly, Purity and Clearness, giving a manifest Splendor in the Fire, but not burning: for after consolidation of the purified parts, it is left to the actual Fire to burn up or consume all extraneous Superfluities not consolidated: wherefore purification is necessary.

XXIII. Sixthly, A fixing Earth, temperate, thin, subtil, fixed, and incombustible, giving permanency of Fixation, in the solution of the Body adherring to it; standing and persevering against the force of the strongest Fire: for immediately after Purification, fixation necessarily follows of course.

XXIV. Seventhly, Tincture White or Red, giving a splendid or perfect Color White, or intensly Citrine, viz, the Lunification or Solification of the Bodies to be transmuted; for that after fixation a pure Tincture or Color tinging another Body; Or a Tincture, tinging the Matter to be transmuted into true Silver or Gold, is absolutely necessary.

CHAPTER LX
OF THE THREE ORDERS OF THE MEDICINE.

1. Of Medicines of the first Order.

I. Subtilty of the matter is necessarily required, as well in the preparation of Bodies, as in the perfecting of the Medicine; because of how much the greater weight, Bodies to be transmuted are, so much greater is the perfection they are brought to by Art; for which reason we shall here declare the differences of all Medicines, which is three fold, according to three Orders.

II. A Medicine of the first Order is every preparation of Minerals, which projected upon the imperfect Bodies, impresses upon them an Alteration, but induces not a sufficient Compleatment; yet the altered Body is thereby changed and Corrupted, with the total evanishing of the Medicine, and all its Impressions.

III. Of this kind is every Sublimation dealbative of Mars or Venus which receives not Fixation: and of this kind, is every additament of the Color of Sol and Luna, or of Venus commixed, and Zyniar, and the like set in a Fornace of Cementation.

IV. This Order changes with a mutation not durable, by

diminishing it self by Exhalation or Evaporation. And of this kind are these described, Chap. 44. Sect. 15, 16, 17. Chap. 45. Sect. 18, 19, 20, 21, 22, 23. and Chap. 46. Sect. 6, 7, 8, 9. aforegoing. And the Work of this first Order is called the lesser Work.

2. Of Medicines of the second Order.

V. A Medicine of the second Order, I call every preparation, which being projected upon Bodies diminished from perfection, alters them to some certain degrees of perfection, wholly leaving other degrees of Corruption, as is the Calcination of Bodies, by which all that is fugitive is burnt away and Consumed.

VI. And of this Order are the Medicines Tinging Luna perpetually yellow, or perpetually dealbating Venus, leaving other differences of Corruption in them.

VII. Now seeing the Medicine of Bodies to be cleansed is one; but of Argent Vive perfectly Coagulable another, we will first of all declare the Medicines for Bodies: and then afterwards the Medicine of the same Argent Vive, coagulable into a true Solisick and Lunisick Body.

VIII. A Medicine of the second Order is that which

does indeed perfect imperfect Bodies, but with one only difference of perfection. But seeing there are many causes of Corruption in every of the imperfect Bodies, as in Saturn a Volatile Sulphureity, fugitive Argent Vive (by both which Corruption must necessarily be induced,) and its Terrestreity: therefore Medicines of this second Order, are such as can only remove one of them, or covering it, adorn the same, leaving behind it, all the other causes of Imperfection.

IX. Since then in Bodies, there is somewhat impermutable, which is innate to them in their Radix, and which cannot be taken away by a Medicine of this Order: that Medicine, which totally removes, that from the mixtion, must be a Medicine of the third and Greater Order.

X. And because we find the Superfluities of things Volatile, to be removed, by way of Calcination; and the Earthiness, not innate, abolished by repeated Reductions; therefore there was a necessity of inventing of a Medicine of this second Order, which might indeed palliate the innate, soften the hard, and harden the soft Bodies, according to the perfection of their Natures, and not Sophistically; but perfectly constitute a true Lunisick, or Solisick, of imperfect Bodies.

XI. Since then it is manifest, that in Bodies only Soft the hastiness of Melting cannot be taken away, by the Artifices of this Work; nor the innate impurity in the Radix of their principles be removed, the Invention of this Medicine was necessary, which in projection might Inspissate their Tenuity, and Inspissating, harden them, to a sufficiency of Ignition with their Melting.

XII. So also in hard Bodies, attenuating their Spissitude, to deduce them to a sufficient Velocity, Liquefaction or Melting, with their own property of Ignition; and palliating them, to adorn the dowdiness of Bodies of either kind, transmuting the one into White, the other into Red most perfect.

XIII. This Medicine is differenced from a Medicine of the third Order, only by Imperfection of a lesser or meaner preparation. But the Medicine Inspissating the Tenuity of soft Bodies, requires one kind of preparation with a Consumptive Fire: and that Attenuating the Spissitude of hard Bodies, another, with conservation of their Humidity: of which kind are those in Chap. 43. Sect. 16, 17, 18, 19, 20, 21. and Chap. 44. Sect. 19, 20, 21, 22. aforegoing, which are in a mean or middle Order.

3. Of Medicines of the third Order.

XIV. This is every preparation, which when it is projected upon Bodies, takes away all Corruption and perfects them, with all the differences or signs of perfection. But this is one only, and therefore by reason of it, we are not obliged to the use of the ten Medicines of the second Order.

XV. Of this Order there is a twofold Medicine, viz. Solar and Lunar, yet but one in Essence, and which have but one way in Operating; and therefore by our Ancestors, whose writings we have read; it is called One only Medicine.

XVI. However there is an addition of a Citrine Color, made of the most clean substance of fixed Sulphur which constitutes the difference between the one for the white, and the other for the yellow, viz, the Lunar and Solar Medicine, the latter containing that Color in itself, but the other not.

XVII. This is called the third Order, or Order of the Greater Work; and that because greater Care, Prudence, and Industry is required in the Administration thereof, and the preparation thereof to perfection, than in any of the former; and also for that it needs greater Labor and longer time to compleat it for the highest Purity.

XVIII. Therefore the Medicine of this Order is not diverse in Essence from the Medicines of the second Order, but only in respect of Degrees, as being more subtilized, and exalted to a much higher degree of Purity, Tincture, and Fixity, in the making and preparation thereof, with a long continued course of Labour.

XIX. All which degrees in their proper place are declared with sincerity of Speech, and the way of preparation Exactly, with its Causes, and manifest Verity; as also the many degrees by which it is brought to Perfection.

XX. For the Lunar Medicine needs one way of preparation: but the Solar another, for the perfect preparation of its Tincture, with the Administration of Sulphur Tinging it: of which we have abundantly Spoken Chap. 46. Sect. 11, 12, 13. Chap. 47. Sect. 11, 11, 13. and Chap. 48. Sect. 43, 44. aforegoing.

CHAPTER LXI

HOW INGRESSION IS PROCURED.

I. Because it happens that a Medicine will sometimes mix, and sometimes not, therefore we shall here declare the way of per—mixing, i.e. how everything, or each particular Medicine not entering, may most profoundly acquire Ingress into a Body.

II. The way is by dissolution of that which Enters, and by dissolution of that which Enters not, and by commixing both Solutions: for it makes everything to be Ingressive, of what kind soever it be, and to be conjoyned through its least parts.

III. Yet this is compleated by Sublution: and Fusion is also, accomplished by the same, in things not otherwise Fusible: whereby they are more apt to have Ingress, and to transmute.

IV. This is the cause why we Calcine some things, which are not of the nature of these, to wit, that they may be the better dissolved: and they are dissolved, that they may the better receive Impression from them; and from them likewise, by these be prepared and cleansed.

V. Or, We give Ingress to these which are not suffered to enter by reason of their Spissitude, or Thickness, with a manifold Repetition of the Sublimation, of Spirits not Inflamable upon them, to wit, of Arsenick, and Argent Vive not fixed; or with manifold Reiteration of the Solution of that which has not Ingress.

VI. Yet this is a good Caution concerning things Impermixable, viz. That the Body be dissolved, which you would have to be changed and altered by these: and the things likewise Dissolved, which you would have both to enter and to alter.

VII. Nevertheless Solution cannot be made of all parts, but of some; with which this or that Body, not another, must be imbibed time after time.

VIII. For by this means it has Ingress only into this or that, necessarily; but this does not necessarily happen into any other Body.

IX. Everything then must needs have Ingress by these ways; by the benefit whereof, it depends on the nature of that, to have Ingress (as we said before) and to Transmute with the Commixtion found out.

X. By this precedent Discourse, is compleated our said

number of Ten Medicines, with a sufficient Production of them, (in order to the Great Work itself.)

CHAPTER LXII
OF THE CINERITIUM.

I. The Solar and Lunar substance is only permanent in the Tryal by the Cineritium: Therefore searching out the true Differences of the Substances of these perfect Bodies and likewise the Causes of the Cineritium, we shall make tryal which of the Imperfect Bodies do more, and which do less endure or abide in the Examen of this Magistery.

II. But we have already sufficiently declared the Secret of these two Bodies in the Profundity of their substance, viz. That their Radix, or first Principle of being, was a large quantity of Argent Vive, and the purest substance of it; at first more Subtil, but afterwards Inspissate, till it could admit Fusion with Ignition.

III. Therefore whatsoever Bodies diminished from Perfection, have more of Earthiness, the less abide or endure in this Examen; but what have less Earthiness, do more endure it.

IV. Because these do indeed more adhere, by reason of the Subtilty of their Parts, closely Permixing and Uniting them: So likewise, Bodies that are of greater Tenuity, or on the contrary, of greater Spissitude, than those which are in Perfection, must necessarily be wholly separated from the Commixtion.

V. For being not of the same Fusion, they are for that cause sake separated: And indeed Bodies which partake of a lesser quantity of Argent Vive, are more easily separated from the said Commixtion.

VI. 'Tis evident then, that seeing Saturn is of much Earthiness, and contains but a small quantity of Argent Vive, and of an easie Tenuity for Liquefaction, which are mostly opposite to a Cineritious Examen; therefore of all Bodies, by the Artifice of the Cineritum, it least endures in the Commixtion, yea it is separated and vanishes most speedily.

VII. Seeing therefore of all imperfect Bodies, it most gives way and receeds; by that it is more fit for the Examen of our Magistery, and the reason is, because it sooner takes its flight, and sooner draws every of the imperfect Bodies with its self from the mixture.

VIII. Also by reason of this, the greater quantity of the perfect Bodies is preserved for the strong

Combustion, or mighty devouring force of the Fire of the Examen: and therefore by the tryal of Lead, it is less burnt, and more easily purified.

IX. And because the substance of Jupiter, consists more of Argent Vive, and partakes of a lesser quantity of Earthiness, whereby it is of greater purity, and of a more subtil substance; therefore it is more safe in the Mixtion, than Saturn and Venus; because it more adheres in the profundity thereof.

X. And for this cause a larger quantity of the perfect Body is absurned, before Jupiter conjoyned can be separated from the Commixtion: Venus gives Fusion with Ignition; but because its Fusion is slower of a perfect Body, therefore it is separated from the Commixtion, yet more slowly than Saturn, by reason of the Ignition of its fusible Substance.

XI. But because it contains less of Argent Vive, and has more of Earthiness, and a more thick Substance, therefore it is more easily separated from the Mixtion than Jupiter, because Jupiter more adheres in the profundity than Venus.

XII. Mars has not Fusion, and therefore is not permixed, which is caused for want of Humidity: but if it happens that it is per-mixed with vehemency of

Fire; then because it has not Humidity enough of its own, by imbibing the Humidity of Sol or Luna, it is united thereto in its least parts.

XIII. Therefore, Tho' it has much Earth, and little Argent Vive, and wants Fusion, yet it can by no flight Artifice be separated from them. By this Artifice (i.e. of the Ciniritium) you come to the true rectification of every Body, if you understand perfectly what we have writ.

XIV. There are two Bodies perfect, abiding this tryal, to wit, Sol and Luna, by reason of their good Composition, which results from their good Mixtion, and the pure Substance of them.

XV. The way of working this Tryal is thus, *Take sifted Ashes or Calx, or Pouder of the Bones of Animals Calcined, or a Commixtion of all, or some of them; moisten with Water, and make the mixture firm and solid with your hands; and in the midst of it, work it into a round flatish lump; make a round and smooth hollowness, and upon the bottom of it strew a small quantity of Glass beaten to Pouder, which lay to dry.*

XVI. *When dry, Put your Metal into the Hollowness thereof, which you would try or prove; put Coals of Fire upon it, and then blow with Bellows upon the*

Surface, till the Metal flows: upon which, being in flux, cast part after part of Lead, and blow with a flame of strong Ignition.

XVII. *Whilst you see it agitated with a strong Concussion, it is not pure; therefore wait till all the Lead, be Exbaled: when that is gon off, and the Motion yet ceases not, it is not yet pure: cast Lead then again upon it, and blow as before, until the Lead vanish. If it do not yet rest, repeat the casting in of more Lead, and blowing upon it, till it be still or quiet, and you see it clean and clear in its Superfices.*

XVIII. *This done, take away the Coals, scatter the Fire, and put Water upon the Test, for you will find it throughly proved: and If while you are blowing this proof, you cast in Glass, the Bodies will be the better and more perfectly purified; because that takes away the Impurities, and separates them.*

XIX. Or, *Instead of Glass, you may cast in* Salt, Borax, *or a little* Alum: This Examen of the *Cineritium* or *Test,* may in like manner be made in a Crucible of Earth, if the fire round about it be blowed, and upon the surface also of the Crucible, that the *Body* to be proved, may the sooner flow, and be perfected.

CHAPTER LXIII
OF CEMENTATION AND ITS CAUSES.

I. We now come to the Examen of *Cement and* whereas some bodies are more, and others less burned by the Calcjnatjon of fire, i.e. they which contain a greater quantity of *Sulphur* (burning) more, but they which contain less, less: Therefore seeing Sol, has a lesser quantity of Sulphur, than other Metallick Bodies, it is not (in the midst of all Mineral Bodies) burnt by the force of fire.

II. And seeing Luna also, next to Sol, partakes of a less quantity of Sulphur, than the other four Bodies; yet has more Sulphur than Sol; therefore it can less bear the strong Ignition of a violent Fire for a long space of time, than Sol can: And by consequence, less bear things burning by a like nature, but Venus less than it, because it consists of more Sulphur still, and of greater Earthiness than Luna, and so can less bear the violent force of Fire.

III. Jupiter also less than Sol or Luna, because it partakes of greater Sulphureity, and Earthiness, than either of them; yet it is less burnt by violence of Fire than Venus, but more than Sol, or Luna.

IV. Saturn in its Commixtion by nature, holds more of Earthiness and Sulphureity, than either of these before named; and therefore is more burnt, by Inflamation or violence of Fire, and is sooner, and more easily inflamed, than all the said Bodies; because it has Sulphureity more nearly conjoyned, and more fixed than Jupiter.

V. Mars is not burnt by itself but by Accident, for when it is mixed with Bodies of much humidity, it imbibes that Humidity, by reason of its own want of the same; and therefore being conjoyned it is neither inflamed nor burned, if the Bodies with which it is joyned or united, be neither Inflamable nor Combustible.

VI. But if Combustible Bodies be mixed with it, it necessarily happens (according to the nature of the Combustion) that Mars is burnt and inflamed. Seeing therefore, that Cement is made of Inflamable things, the necessary cause of its Invention is manifest, viz, that all Combustible things might be burned.

VII. And since there is but one only body incombustible, that alone, or what is prepared according to the nature of it, is kept safe in Cement. But which abide more, and which less, are known with their Causes, Luna abides more, but Mars less, Jupiter yet less, and Venus less than Jupiter, but Saturn least of all.

VIII. The way of Examination by Cement is thus. *You must compound it of Inflamable things, of which kind are all blackening, flying penetrating things, viz. Vitriol, Sal Armoniack, Verdigrise, Alum, or Plumous Alum, and a very small quantity of Sulphur, with Humane Urine, and other like acute, and penetrating things: All which are made into a Paste, with the Urine aforesaid, and spread upon thin plates of that Body which you intend to examine by this way of Probation.*

IX. *Then the said plates must be laid upon a Grate of Iron, included in an Earthen Vessel; but so as not to touch one another, that the power of the Fire may have free and equal access to them. Thus the whole must be kept in Fire, in a strong Earthen Vessel for the space of 3 days, but with this Caution, That the plates may be kept Red Fire hot, but not melt.*

X. After the third day, you will find the Plates cleansed from all impurity, if the Body of them was perfect; if not, they will be wholly corrupted and burnt in the Calcination.

XI. Some expose Plates of Metal to Calcination,

without a Composition of Cement and they are purified in like manner, if the Body be perfect: If not, they are totally consumed: But in this kind of Examen they must have a longer space of time, (for that they are purified by the only force of Fire) than if they were Examined by the help of Cement.

XII. And for that the nature of Luna differs not much from the nature of Sol, therefore of necessity it rests with it in the Tryal by Cement, and there is no separation of Bodies one from another in these two kinds of Tryal, unless that be caused by reason of the Diversity of the Composition of their substances.

XIII. For from thence results the Diversity of Fusion, and Thickness, or Thinness or Rarity, which are indeed the causes of Separation; for that, by reason of the strong Composition of some, their substance is not corrupted by the substance of the Extraneous Body, in as much as a mixtion of them, cannot be made through their least parts.

XIV. Therefore in such a commixture, they must necessarily be separated each from other, without the total corruption of their Essences. And the perfecting of imperfect Bodies is discerned, when they are by Ingenuity of preparation found to be of the same Fusion, Ignition, and Solidity.

CHAPTER LXIV
THE EXAMEN BY IGNITIONS

I.   Since Bodies of greatest Perfection, with determinate Ignition, are found to receive the Fire before fusion of them; therefore we say, if our design is to find out the compleat alteration of them, there is a necessity to bring such Bodies to their Fusion.

II.   And before these perfect Bodies be Fused, we see them admit Ignition with Inflamation of a pleasing Celestine Color, and this before their Ignition comes to the whiteness of Fire, which by the Eye can in in no wise be discerned.

III. 'Tis evident then, that the perfect Ignition of them is before Fusion, with intense Redness, and not with whiteness, which the Eye cannot behold; for if the prepared Bodies be Melted, before they are red hot with Fire, they stand not in perfection.

IV. And if they be made Red Fire hot with labor, and great Violence of Fire, their preparation is not true and perfect; and this indeed if it happen in soft Bodies, for that the same is only found in Mars.

V.   Because Ignible Bodies do not easily in the way

of preparation admit Ignition; nor Fusible Bodies the right Fusion, which we find to be in Bodies perfect according to Nature.

VI. If Bodies prepared, in their Ignition, give not a flame of a pleasing Celestine Color, their preparation is not compleat.

VII. And if any part of the Weight, Color, Beauty, Ignition and the like, be found diminished, by reason of the Differences, or force of the Preparation, you have not rightly proceeded: therefore you must search again till you find out your Error, and chance to hit upon the right way through the Divine goodness.

CHAPTER LXV

THE EXAMEN BY FUSION OR MELTING.

I.   Fusion with Ignition is the only Argument of Perfection; yet not with every kind of Ignition, but with Ignition in which the Body waxeth not altogether white; and with Ignition in which is not made a dull paleness of Fire, and in which, the body is not suddenly Melted, or flowes not immediately after Ignition.

II. For when a body flows with the very small force of a weak Fire, either without Ignition, or with a pallid Ignition; the body thus prepared, must needs be still an imperfect body.

III. And if a body after Fusion, be not suffered presently to coole, and its Ignition be presently turned wholly into blackness, and by reason thereof, looseth its Ignition, before it becomes hard, it is not a body brought to perfection, of what kind forever it be. Now this is from its softness, and is one of the kinds of imperfect bodies.

IV. If the Ignition of a body before Fusion thereof be made with great Labor, and Violence of a strong Fire, and with a Ray of brightness Inestimable, altogether white and shining, it is not a perfect body, but a body of hardness altered.

V. If also after Fusion thereof, and when taken from the Fire, it be presently hardned, that it flows not, the fulgent Ignition thereof yet remaining, it is not a body of Lunar or Solar perfection, but comes under the nature of the differences of Mars.

VI. By what has been said, then, it is evident, that in bodies Fusible, a three-fold Ignition may be found before Melting of their Substances, viz, one Pallid; another Red and clear; and a third most white, shining with Rays.

VII. The first of these is an Ignition of soft Bodies; the second of perfect bodies, the third of hard bodies, as is proved by Reason and Experience.

VIII. If you would find out the Degree of all these Ignitions, to compleat all Fusible bodies, you must learn the Compleat sufficiency for the perfection of Fusion; and by considering, recollect, the difference of all the Signs of the Degree of Fusion; thus may you find it out, otherwise not.

CHAPTER LXVI

THE EXAMEN BY VAPORS OF ACUTE THINGS.

I. Perfect Bodies exposed over the Vapors of acute things, viz. things Sharp, Sowre, and Saline, are apt either little or nothing at all to slower, or to emit a most pleasant Celestine Flos.

II. But Sol or Gold flowers not: yet Sol or Luna not pure, being exposed over the Vapors of the said acute things we find to Flower, and to yield a most delectable Celestine Flos: of which, that of Sol is more delightful than that of Luna.

III. We then (from seeing this) imitating Nature, do in manner produce a Celestine Color in prepared bodies, which Color is perfected by the goodness of Argent Vive, as we have formerly declared.

IV. Whatever prepared bodies then, being put over the Vapors of acute things, do not produce a pleasant Celestine Color, they are not yet brought to the total Perfection of their Preparation.

V. There are some bodies, which in the Examen of Saline things, flower in their Superfices, with a dull Red, or dull Citrine Color mixt with Greenness: of this kind is Mars.

VI. Some flower with a dull Greenness, mixt with a Turbid Celestine Color; of this kind is Venus. Some are found to yield a dull white, and of this kind is Saturn: And some a clear White, of which kind is Jupiter.

VII. Hence it is evident that the most perfect Body flowers least, or nothing at all; and if it yields any Flos, it is in a long space of time. And indeed among imperfect Bodies, the Gummosity of Jupiter most slowly admits any Flowers; whence by the Exainen of this Magistery, we find Jupiter in the work of the greater Order, more nearly approximate to perfection.

VIII. And by this Tryal or probation, it may be known, in what kind of temperament, the proposed Body does consist; if you rightly conceive the Order of these things here declared.

CHAPTER LXVII

THE EXAMEN BY EXTINCTION OF BODIES RED FIRE-HOT.

I.   If the Body heat red Fire hot be extinguished in Liquor, and the Lunar yield not a white Color, and the Solar a bright Citrine, but is changed into a Foreign Color, the Body is not, transmuted into the perfection of a perfect Body.

II.   And if in repeating its Ignition and Extinction in the Waters of Salts or Alums, by whatsoever kind of preparation, it yields, a Scoria, of Affinity to Blackness in its Supersices; Or, if in the Extinction of it in Sulphurs, and from the Extinction with often repeated Ignition it vanishes or infects itself with a fould Blackness, or by force of the Hammer breaks into peices, the Work is not perfect.

III. Or, if it with Cementation of the mixture of Sal armoniack, Veridgrise, and Urine, or things of like Nature, be exposed to the Fire, and after the Ignition and Extinction of it (whether Lunar or Solar, it totally looses its proper Color, or makes a Scoria, it is evident, that the Body does remain in imperfection.

IV.   And this we farther give you, as one certain

general Rule, that as well in these present Examens, or Probations, as in the three Exámens following; if among the differences of perfection, the altered or changed Body shall change anything of its weight or color from those of perfection, (and which it ought not to do) you have erred in your Work, and the alteration or change made, is a thing of no good, or profit, but destructive and of disadvantage rather.

V.   There remains yet three other ways of Examination, as appears by Chap. 49. Sect. 7 aforegoing, which should here immediately follow, but that they are treated off in the Chapters, under their several, and respective Titles, viz. The Examen by Admixtion of burning Sulphur, in Chap. 38. Sect. 6, 7, and 8. The Examen by Calcination and Reduction, in Chap. 5; Sect. 32. The Examen by the easie susception of Argent Vive, in Chap. 48. Sect. 38. where the matter is explained at large, and to which we refer you.

CHAPTER LXVIII

A RECAPITULATION OF THE WHOLE ART.

I.   Having now handled the Experiences and Causes of the power of this our Magistery, according to the necessity, order and method of our proposed Discourse, it only remains, that we should at once declare the compleating of this whole Divine Work; and in few words contract the dispersed Magistery into one Sum, in general heads.

II. We say then, that the Sum of the whole Art, and of the Operations of this whole Work, is no other, than that the Stone, Magistery, Elixir, or Tincture (declared in its Chapters) should be taken, and with diligent Labour and Industry, that Sublimation of the first degree be repeated upon it: for by this it will be cleansed from corrupting Impurity.

III. And the perfection of Sublimation, is the Subtilization of the Stone by it, until it can be brought to the ultimate purity of Subtilty, and lastly be made volatile.

IV. This being done, by the way of Fixation, it must be fixed, until it can dwell and remain in the highest Violence or Force of Fire: and herein consists the measure of the second degree of preparation.

V.   The Stone is likewise prepared in the third degree, which consists in the Ultimate compleating of the work, or perfection of the preparation, which is this: The now fixed Stone, you must make by the way of Sublimation Volatile, and the Volatile fixed.

VI.   The fixed you must also dissolve, and the dissolved again make Volatile; and the Volatile again make fixed, until it flow and alter or change into Solisick or Lunisick with all the signs of perfection.

VII. From the reiteration of the preparations of this third degree, results the Multiplication of the Virtue and Quantity of the Medicine in goodness and purity to the highest perfection in kind.

VIII. From the diversity then of the Operations reiterated upon the Stone, Elixir, or Tincture, in its degrees, results the variety of the Multiplication of the goodness of the Alteration, and quantity of the Medicine for Transmutation according to their kind.

IX.   So that among the Medicines, some transmute into a true Lunisick Body of perfection, and some into a true Solisick Body of the perfection of the Solar kind.

X.   And of these Medicines, some transmute an hundred–fold, as much as their own weight is, some two hundred fold, some three hundred fold, some a thousand fold, and some to infinity, so that from hence it may easily be known whether the magistery is brought up to perfection or not.

XI.   Now that the Envious may not Calumniate us, we declare, that we have not treated of this our Art with a continued Series of Discourse, but have dispersed it in divers Chapters: and this was done, that evil men might not usurp it unworthily: Therefore we have concealed it in its places, where yet we indeed speak openly, and not under an Aenigma, but in a clear and plain Discourse.

XII. Therefore let not the Sons of Doctrine despair, for if they seek it, they may find the same, tho' he who seeks it, following Books only, will very slowly attain to this most desirable Art. As for us, we have described it in such a way of speaking as it submissive to the Will of the Most High, Blessed, and Glorious God, writing the same as it chanced to be recollected, or was infused, by the Grace of his Divine Goodness, who gives it to whom he pleases, and withholds it from the Foolish and Unworthy.

Here is the Sum and the end of all G E B E R' S Works.

F I N I S.

A Word from the Publisher

Thank you for purchasing this small work from The R.A.M.S. Library of Alchemy. During his lifetime, Hans Nintzel was dedicated to the identification, acquisition, study, retyping and, when necessary, translation of what he considered to be the most important known works on Alchemy. Hans was assisted by his sparse network of fellow Alchemists, all members of the Restorers of Alchemical Manuscripts Society (R.A.M.S.). I was an active member of R.A.M.S.

My goal is to publish all of the works originally made available through R.A.M.S. as photocopies. To facilitate this, I have chosen to have the books professionally printed. I also have a few titles that I intend to add to the original R.A.M.S. Library, selected by strict criteria established by Hans.

If you have a work on Alchemy that you believe should be a part of the R.A.M.S. Library, please contact me through R.A.M.S. Publishing Company.

Philip N. Wheeler

www.ingramcontent.com/pod-product-compliance
Lightning Source LLC
Chambersburg PA
CBHW080804180526
45168CB00006B/2329